Experiential Spectatorship

Experiential Spectatorship offers a lens for analyzing audience experience with(in) a variety of contemporary media. Using a broad-based perspective, this media includes participatory theatre, video games, digital simulations, social media platforms, alternate reality games, choose your own adventure narratives, interactive television, and a variety of other experiential performance events. Through a taxonomy that includes *Immersion*, *Participation*, *Game Play*, and *Role Play* the book guides the reader to understand the ways mediatization and technics brought about by digital technologies are changing the capacities and expectations of contemporary audiences. In their daily interactions and relations with their technologies, they become mediatized spectators. By reading these technologies' impacts on individual subjectivity prior to acts of spectatorship, one gains the tools to best describe how the spectator creates forms of relational exchange with their experiential media.

This book prepares the reader to think in a digital manner so they can best recognize how performance and spectatorship in the twenty-first century are evolving to meet the needs of future waves of spectators brought up in a postdigital world.

William W. Lewis is Assistant Professor of Theatre History, Literature, and Criticism at Purdue University. He is an interdisciplinary scholar/artist whose work focuses on the fields of intermedial, postdramatic, and devised performance. His research explores the role of mediatization on contemporary audiences and the implications of constant connection to media on how we teach theatre-making.

Audience Research

Series Editor: Kirsty Sedgman, Kelsey Jacobson

The Routledge Theatre & Performance Series in Audience Research is an interdisciplinary forum for investigating audience response and spectatorship via empirical methodologies.

The primary aim of this series is to give scholars a dedicated place not simply to disseminate the findings of their empirical research, but to reflect on the relationship between findings and methodology: more specifically, between the specific ethical and practical features of methodological design, the information they have uncovered, and the knowledge they have produced as a result.

This series offers a vital home for this fast-growing field.

Routledge Companion to Audiences and the Performing Arts
Matthew Reason, Lynne Conner, Katya Johanson, Ben Walmsley

Impacting Theatre Audiences
Methods for Studying Change
Dani Snyder-Young and Matt Omasta

Audience Experience and Contemporary Classical Music
Negotiating the Experimental and the Accessible in a High Art Subculture
Gina Emerson

Experiential Spectatorship
Immersion, Participation, and Play During Times of Deep Mediatization
William W. Lewis

For more information about this series, please visit: *https://www.routledge.com/Audience-Research/book-series/SAR*

The Mediatized Spectator

Experiential Spectatorship

Immersion, Participation,
and Play During Times of Deep
Mediatization

William W. Lewis

Routledge
Taylor & Francis Group

LONDON AND NEW YORK

First published 2025
by Routledge
4 Park Square, Milton Park, Abingdon, Oxon OX14 4RN

and by Routledge
605 Third Avenue, New York, NY 10158

Routledge is an imprint of the Taylor & Francis Group, an informa business

British Library Cataloguing-in-Publication Data
A catalogue record for this book is available from the British Library

ISBN: 978-1-032-50289-2 (hbk)
ISBN: 978-1-032-50507-7 (pbk)
ISBN: 978-1-003-39881-3 (ebk)

DOI: 10.4324/9781003398813

Typeset in Sabon
by KnowledgeWorks Global Ltd.

Contents

Figures

Acknowledgements

This book has been in progress for over a decade and was influenced by many people I owe many thanks to. It began as an inquiry into the ways digital tools offer performance makers avenues for engaging audiences for social change. The first major shift came when I was introduced to the work of Blast Theory by Dr. Amma Ghartey-Tagoe Kootin. She inspired me to broaden my perspective and to look outside of the conventional theatrical frame. Without her all my work would be significantly different. Along the path many others shaped my ways of looking at the world which seeps into the pages of this book. I must thank Sarah Bay-Cheng for her mentorship and feedback while working on the earliest versions of this project; the members of the IFTR Intermediality in Theatre and Performance working group for their generous insights on the intersection of digital culture(s) and performance, with a particular shout out to Liam Jarvis, Karen Savage, Ralf Remshardt, and Rosemary Klich; to Beth Osnes, Oliver Gerland, Bud Coleman, Marcos Steuernagel, and Claudia Orenstein for supporting my ambitions as an interdisciplinary scholar-artist while in grad school; Sarah Johnson and Sean Bartley for their peer feedback, collaboration, and friendship; Matt Adams, Nick Tandavanitj, and Ju Row Farr of Blast Theory for access to their works and processes; and most importantly to my partner Benji who has been my personal support system, theoretical sounding board, and guide to all things important in life.

I also want to thank the editors and publishers at *Performance Research*, *Theatre Topics*, and the *Journal of Dramatic Theory and Criticism* where parts of this book were first published in earlier forms.

These articles include:

William W. Lewis. 2017. "Performing 'Posthuman' Spectatorship: Digital Proximity and Variable Agencies." *Performance Research* 22, no. 3: 7–14.

William W. Lewis and Sarah Johnson. 2017. "Theatrical Reception and the Formation of Twenty-first-Century Perception: A Case Study for the iGeneration." *Theatre Topics* 27, no. 2: 123–136.

William W. Lewis. 2023. "Resisting Algorithmic Determination: Becoming the Political Other in Blast Theory's *Operation Black Antler*." *Journal of Dramatic Theory and Criticism* 37, no. 2: 49–72.

Introduction

Back in 2012, I attended a conference roundtable on new media and theatre focusing on the possibilities of interdisciplinarity. The panel intended to discuss how theatre departments should manage the integration of approaches from film, media studies, and digital culture. During the question-and-answer session, an attendee offered an anecdote illustrating a way of thinking about a nearing future subsumed by all things digital. His story encouraged a novel outlook on the inclusion of both digital humanities methods in education and digital media in theatre, one in which theatre practitioners, critics, and educators fluidly adapt to the pervasive nature of contemporary digital technologies. He asked the room to make a gesture for how they would make something bigger. Nearly everyone raised their hands and widened their reach outward by roughly four feet. He then explained how his three-year-old niece responded when asked the same question: "She raised one hand slightly in a pinched gesture and moved her thumb and forefinger outward about two inches."[1] Based on this response, it was clear that this child's worldview and sense of being-in-the-world were deeply affected by their digital devices. This child embodied the emergence of a growing wave of cultural perception,[2] which informs the ways technologically embedded societies read, watch, and perform. Spectators belonging to this child's generation will perceive the world differently than those born before the mass adoption of the personal computer, internet, smartphone, and tablet. Spectators from generations before hers first experienced the world and developed their perceptual apparatus somewhat removed from the digital influences most are steeped in today.

The full social impact of technologies does not fully emerge until they have become commonplace: "It's when a technology becomes normal, then ubiquitous, and finally so pervasive as to be invisible, that the really profound changes happen, and for young people today, our new social tools have passed normal and are heading to ubiquitous, and invisible is coming" (Shirky 2008, 105). As a member of the last years of Generation X, I grew up in an era where an explosion of communications technologies rapidly changed the way the world around me worked. The home computer,

DOI: 10.4324/9781003398813-1

internet, cellphone, and smartphones went from novelties to common place necessities. I was born into an era where these devices didn't help condition my subjective experience of the world from early development though. My childhood primarily consisted of play, learning, and exploration using my imagination connected to non-digital worlds. While the television was a dominant mode of mediation, it was relegated to the living room setting and not the constant and pervasive presence it has become when remediated into digital formats. This meant much of my early days were spent outside on my BMX bicycle, digging in the dirt with GI Joe action figures, lugging around backpacks full of heavy books, chalk dust on my hands after class, wandering in the woods trying to find the end of a stream, and chasing dogs in the neighborhoods in which I lived. While I regularly watched Saturday morning cartoons and the occasional Nickelodeon rerun, it wasn't until I was ten that I had the luxury of enjoying a video game away from the arcade. My first personal computer didn't arrive until age 20. I list these childhood exploits not to indulge in wistful nostalgia but rather to show how different it was to grow up divorced from constant connection to digital realms.

Those coming of age today were raised in digital cultures whose primary media source is the internet, delivered via multiple pervasive interfaces. To better understand how to bridge the cultural and subjective digital divide that primarily exists between many established scholars and educators in our field and these future audience members, new lenses for understanding how contemporary spectators perform through their connections to technology are important. As I'll argue throughout this project, when performing as spectators, these mediatized subjects increasingly crave constant interaction that matches their evolved ability to multitask with the media that has become commonplace. Their connection to interactive technologies is ushering forth cognitive, social, and cultural evolutions through mediatization. Discussing these evolutions leads to a better understanding of how to communicate with coming waves of spectators who have formulated their worldview in an era where digital technologies are seamlessly integrated into everyday life and are therefore invisible. Throughout this project, I discuss a process of evolution using the term technogenesis: Defined as the co-evolution of the human perceptual apparatus with its technological environment. By better understanding the relationship between mediatization and technogenesis and their impact on spectatorship, a valuable analytical tool becomes available to help bridge the gap between those who were not born into a paradigm of pervasive and invisible digital technologies and those who were. The incorporation of short spectator experiences, which I describe as *Instances*, helps to make visible the invisible operations of contemporary technologies through the staging effects of experiential spectatorship.

The speed of twenty-first century technological advancements is upending long held traditions within social paradigms globally. Within the interconnected fields of performance and media writ large, these advancements are causing a necessary need to re-evaluate the relationship between the audience and the performance frame through which individuals perform their role as spectator. This project is in service of that re-evaluation, intent on putting into relation the spectator with the ongoing impacts of technology and media on the social sphere. To best understand how the combined forces of technology and media work on the social, it is useful to turn to mediatization theory. Mediatization is a theoretical construct used to describe the way communication technologies have spread across time and geographic areas to encompass all aspects of both public and social life, which in turn creates a symbiotic relationship between media and social systems (Eskjaer 2018; Hjarvard 2013; Lundby 2009). As a process, mediatization is most concerned with the relations between media and other social domains and specifically how these domains are changed and then bring about new conditions for communication and interaction among human and non-human systems (Hjarvard 2018).

For clarity, let me begin by explaining that this project will define all frames of/for performance as media so as not to create a hierarchical dynamic of tension between each. Performance is considered as "the medium through which we perceive a series of interrelated events" (Bay-Cheng 2012, 35). This definition allows us to consider theatre, dance, film, television, video games, social media platforms, opera, museums that contain art and exhibits, and a variety of "staged" events as media. This formulation of media refers to any aesthetic frame for the transmission of information using communicative means. This is similar to how the word medium could refer to a container for holding information/matter that is intended to be transferred or exchanged. That exchange becomes performance when the spectator is put into relation with the media. Of particular interest to this project is media that create an interactive and reciprocal relationship between the spectator and the media container. This relationship is at the center of all experiential performance events. These are mediated events where the spectator's experience becomes a crucial interactive and relational component of the overall structure of the event. Various modes of experience can then be categorized and analyzed based on formal characteristics of exchange between the spectator and event. Due to the influence of mediatization, however, these spectatorial acts should first be considered inherently mediatized, as the spectator themself has been mediatized. They operate as a ***mediatized spectator.***

While mediatization should not be considered a single overarching determinate of social life, it is a powerful one that has become increasingly intertwined with virtually all social systems in the twenty-first century, including politics, culture, religion, and economics. Through mediatization, a certain set of media logics become important to understand if one is to analyze any human act within the social sphere. This project largely relies on the theorization of mediatization offered by Nick Couldry and Andreas Hepp (2017) in their book *The Mediated Construction of Reality* which is situated as a key text in the emerging fields of media sociology and media-focused materialist phenomenology. Mediatization is inherently a social process and cannot be understood divorced from the communicative actions of the individual who collectively becomes a part of and shapes the social spheres they live within. The authors argue for a materialist phenomenological perspective that considers the interaction between the individual and their environment, including all the various media material and technological affects. It is in this environment that "the very elements and building blocks from within a sense of the social is constructed become themselves based in technologically based processes of mediation" (7). The framework they offer, along with others on mediatization helps to set the stage, so to speak, for a better understanding of how spectatorship operates within a paradigm of deep mediatization. Deep mediatization refers to the current paradigm where processes of mediatization can no longer be divorced from the social (Hepp 2019). This correlates with Matthew Causey's (2016) explanation of the postdigital paradigm, where digital processes and cultures can no longer be thought of as separate from contemporary culture, requiring us—as artists/scholars/educators of media/performance—to develop the tools to think digitally. Thinking digitally is something we all do naturally but often do not do so purposefully. For those of us in theatre and performance, it is critical to harness this way of thinking. As Sarah Bay-Cheng (2023) so clearly states about our present moment,

> By engaging more thoroughly with media theory and digital methodologies, performance and theatre studies is perhaps uniquely positioned to take hold of this current moment, not only to understand it, but also to communicate that understanding in ways that will create more equitable and sustainable futures. (21)

This engagement with media and thinking through the digital is at the heart of this project. By adopting the epistemologies of mediatization and the mediated, we gain a better grasp on the multiple ways the spectator operates in a deeply mediatized and postdigital world.

The Critical Frame of Mediatization and Technogenesis

Mediatization operates as a meta-process where human communication is mediated by technological paradigms which, in turn, fundamentally impacts the nature of reality via the social as a structural component of human life. Couldry and Hepp argue we have entered an age of deep mediatization, where the social is constantly evolving in flux based on the exchange between subjects (humans) and objects (communicative media). They identify four waves of mediatization that include *mechanization, electrification, digitalization,* and *datafication.* As this project is primarily concerned with contemporary spectatorship, which has become a key area of inquiry in the last 20 years, and specifically spectatorial acts with(in) experiential frames, the scope is narrowed to the experiences of mediatized spectators shaped by processes of digitalization and datafication. A more defined description of these waves and their relationship to mediatization is covered in the following chapter where mediatization is connected to another important technological process described as technogenesis.

Technogenesis operates as a co-evolutionary process between individual human beings and the technologies that shape their perception of the world in which they live (Hayles 2012; Stiegler 1998). It is embedded in the very fabric of human social systems and is fundamentally a relational process between human beings, technologies, and the environments (social, political, ethical, moral, etc.) that contain, interact, and inform humans and technologies from the very beginning of human history. Understanding the process of technogenesis as one built out of ongoing technological relations helps to deconstruct the often-held notion of evolution as a process based in linear causality. Instead, technogenesis operates as a form of evolutionary adaptation that is continually in flux based on the technological environments that humans operate with(in). Technologies are what allow media to form into specific containers for communication. In this sense, technology refers to practically any tool that allows people to either make meaning out of the world and/or manipulate the world around them. In today's deeply mediatized society, there are a plethora of overlapping and intricately interconnected technological tools and processes shaping human perception. When expanding out beyond individual tools and technologies, we can think about how communication is shaped through various media, which are uniquely entangled with today's technologies. For instance, the television is both a technology and a medium for the transmission of communicative meaning-making using a combination of image and sound.

By pairing technogenesis with mediatization, I intend to show how technology's capacity to reshape human perception correlates with evolving modes of experiential performance becoming increasingly prevalent in societies that operate under deep mediatization. To argue this, I offer a historical reading of the introduction and growing pervasiveness of technological affects, which I refer to as *technics* that arise within the above waves as meso-processes of technogenesis. These technics are then offered as analytical tools for both understanding the rise in experiential forms of performance and the relational acts of exchange that occur between spectators with(in) these experiential forms. These forms are broken down into a four-part taxonomy which are read alongside individual technics. By describing how technogenesis impacts the subjective experience of reality via mediatized sociality, we can begin to better explain the shifts toward increasing expectations for performance interactivity across our media ecologies. This taxonomy is not offered as a way of simply further breaking down modes of performance, that has largely already been done. Instead, it is offered to work through the ways spectatorship operates in a broadly interdisciplinary mode across various fields of communicative media.

This project introduces a framework for examining acts of spectatorship beyond the ocular focus often dominating scholarship of the twentieth century to expand into multiple forms of reciprocal perception and action. The analytical frame works as a toolset for understanding and discussing acts of spectatorship across these media. This frame lets one see how these forms have become mediatized due to the social configurations and processes impacting the spectators involved. The various media are understood as entry points into analyzing contemporary spectatorship as an evolving and overlapping set of experiential exchanges. By drawing intersecting lines between the historical processes of mediatization, the theoretical processes of technogenesis, and practical acts of experiential exchange, new knowledge emerges concerning the way our connections and relationships with technologies change the way spectatorship operates in the twenty-first century. I propose that a better understanding of these connections and relationships leads to an analytical framework for discussing contemporary spectatorship practices that are based on relational exchange and operate in networks of reciprocal action I discuss as *experiential spectatorship.*

This framework is developed by first breaking down the societal relationship between people and their media/technological environments, then clarifying how these environments shape individual intersubjectivity, which can then be applied to specific acts of spectating. At the level of the individual, mediatization works through the process of technogenesis to impact the subjective position human beings perform. A central focus is on a form of techno-subjectivity connected to a mediatized condition in which societal interaction becomes more complex through the increased visibility of simultaneous interconnections brought about by the invisibility of integrated technologies. Through this mediatized subjectivity, the invisible agency of technology becomes recognizable and crucial to consider, allowing one to discuss how agential elements are folded in that upend the historical binary of performer and spectator, creating a complex relational configuration between technologies and people. By adopting this subjectivity into a critical lens, acts of spectatorship can then be viewed through a mediatized perspective. This perspective looks at the overall condition of mediatization on social systems, how technogenesis works on individuals to change their perceptive capacities with(in) performance configurations, and the ethical considerations of twenty-first century life.

Following the trend to utilize reader-response theory in theatre, performance, and media, spectatorship has often been considered a form of reading (Barker 2006; Bennett 1997; Michelle 2007), in that reading is a form of interpretation through an often ocular and unidirectional form of receiving and comprehending information. This follows a text-centric formulation of reception that helps inform meaning-making. Taking up more recent attempts (Alston 2016; Freshwater 2009; Livingstone 2003; Napoli 2012; Reason 2010; Reason et al. 2022; Reason and Sedgman 2015; Sedgman 2016; 2018; 2019; Walmsley 2019) to shift away from this narrowly defined perspective, I discuss experiential spectatorship as an embodied form of "reading," where one interprets an environment through processes beyond reception-based meaning-making. The human perceptual apparatus operates as the primary medium through which this "reading" happens, and therefore, refers to a process of embodied perception as environmental and social interpretation predicated on feedback cycles of reciprocal exchange. This process works as a sort of "spectating system" (McConachie 2015), where the experience of spectating is considered a blending of emotion, cognition, affective sensing, and action. This "system" becomes the focus instead of simple meaning-making or as Sedgman (2018) explains, "The object of study here is not meaning but the pathways that bring people to those meanings" (315). Since contemporary technologies change the way people read and interpret their environments (Hayles 2012), they also change the way one performs the role of spectator within this system. The spectator's evolving function of techno-environmentally driven subjectivity matches these technologies to be more encompassing of varying levels of embodied perception, requiring a more visibly active response. Following Matthew Reason (2010), we then should focus on "the spectator experience [which] is a perceptual and imaginative doing, a cognitive act which is often accompanied by awareness of the act of cognition. Spectatorship, in other words, is a form of active perception, where we are often (but not always) aware of ourselves looking" (20).

Through deep mediatization, spectatorship becomes increasingly interactive and relational, ushering forth a new way of thinking about performance outside of the more unidirectional model of reception studies presented in previous scholarship on theatrical audiences (Bennett 1997; Holub 1984; Wilson 2008). After all, to receive indicates a one-way mode of information transfer. Experiential spectatorship is instead reciprocal and in perpetual motion due to the constantly in flux process of mediatization. The media logics prompting deep mediatization urge people to navigate the liminal spaces of the actual and the virtual simultaneously, potentially ushering forth the elimination of the mark between the two. This argument is supported by engaging with theories from performance, media, and literature (Castells 2000, 2004; Causey 2006, 2016; Couldry and Hepp 2017; Farman 2012, 2021; Hansen 2004, 2006, 2015; Hayles 1999, 2012; Hjarvard 2013; Jenkins 2006; Krotz 2008) as the ideas that lead us beyond the previous models above. This leads me to the primary question I work through in this project: How do acts of spectatorship in various architectures of experiential performance correlate with changes in subjectivity, communication, and sociality brought about by deep mediatization and technogenesis?

The analytical model combines multiple processes of inquiry informed by current techno-social paradigms of mediatization. They interconnect and weave together in many combinations, though they are often understood and analyzed in a discipline-specific manner. The project is structured in a way that dissects individual practices and processes to allow readers multiple ways to interpret the connections. I describe these processes as *architectures of exchange* which include *Immersion, Participation, Game Play,* and *Role Play* as distinct yet interconnected frames of spectatorial experience.

The term architecture is used as a way of describing a system that encompasses particular sets of practices and processes. An architecture serves as both a framing mechanism or structure in and through which the practices unfold but also acts as the rule sets for how a spectator might perform with(in) the frame. This is similar to how the social processes of mediatization both frame and inform ways of being in the world. The language of architectures comes from the established terminology of system or process architectures, which are often applied and used within computer systems, business structures, and political processes. An architecture, in this mode, is a conceptual model used to help frame a complex set of processes divided up into relational actions of the infrastructure (the inside of the architecture, its individual components, and the interactions made by these components), the suprastructure (outer world that governs the architecture itself and which allows the architecture to become a component of a larger infrastructure), and the superstructure (the mediating structure marking the limen between infra and supra). In a process- or system-based architecture, the inside, the outside, and the mediator are constantly shifting, allowing for change and interaction in multiple directions at once. Using this language allows for the creation of conceptual models where one can focus on the flux of the infrastructure while still understanding the unending impact of the suprastructure.

Couldry and Hepp (2017) describe how social systems are constructed similarly to my use of architectures. The social world exists and can be understood in three levels of abstraction (17–21). At the suprastructure or meta level, the social world is one that operates intersubjectively. It is a world that exists both in relation to actors with agency and also outside of these actors. Considering a human being as one of these actors, the social world is both dependent and independent of that actor at the same time. Its independence comes from the fact that even without the individual actor, it still exists, however, its dependence comes from the fact that without the actor there could be no perception (and hence meaning) of the social world. At the level of superstructure or meso level, change occurs in the social world. It is through the bodies of human beings and the mediating function of technologies that this change can occur. The bodies allow the technologies to exist, and the technologies allow the bodies to act in the world (19). At the infrastructure or micro level, the authors divide up the social world into "internally differentiated domains" (19). These domains serve as the individuated spheres through which communication operates, which, in turn, influence the construction of social reality. However, these domains are not solely independent. They exist in a constant state of friction and overlap with other domains, individual actors, and the entirety of the social world structure. Therefore, the social world is "differentiated into many domains of meaning, even though it is bound together by multiple relations of interdependence and constraint" (20). The entire framework of experiential spectatorship operates in the same way social worlds do.

An Interdisciplinary Relational Model for Audience Interaction

This project draws from multiple disciplines and works at the intersection where Theatre & Performance Studies, Media Studies, and Technology Studies meet, to promote a way of breaking down barriers between them. The ideal audience are those who seek interdisciplinary methods to consider each specialization as part of a larger field. Perspectives on sociological constructions of human life in digitally mediatized society, critical theories from theatre and performance studies, alongside histories of technology, are fused to show how contemporary technologies and media shape us just as much as we shape them. For example, the use of smartphones allows a user the ability to connect to any place and any time at any moment. That connection changes how that user understands space, place, and time as an individual subject prior to acts of spectatorship. A broad-scale approach is offered to better understand the reciprocal relationships between technology and individuals who then utilize media logics to stage experience. Previous work on the relationships between technologies and performance serves as the entry point into these experiences.

Due to the multiple interconnecting strands of methods and theories, there is no one definitive literature review, instead, each section includes a narrowed theoretical approach informed by mediatization. As my primary home is in theatre and performance studies, with a particular focus on inter-, trans-, and multimedial performance, there is often a focus on previous conversations in these sub-fields of inquiry. Each has been uniquely interdisciplinary, putting into conversation the multiple disciplines discussed above. My argument concerning technogenesis and spectatorship could not exist without these foundations. Surveys of technological implementation in performance (Dixon 2007; Salter 2010) are crucial to my understanding of the historical connections between technology and performance. Likewise, the theoretical conversations on media and performance from scholars of digital culture and performance (Auslander 2021; Bay-Cheng 2007, 2012, 2014, 2016, 2017; Bay-Cheng, Parker-Starbuck and Saltz 2015; Benford and Giannachi 2011; Birringer 2000, 2008, 1998, 2006; Broadhurst 2007; Broadhurst and Machon 2006; Crossley 2019; Giannachi 2004; Jarvis 2019; Klich and Scheer 2012) inform my understanding of performance work that is multi-medial. This work also informs my argument for classifying multiple types of performance under the category media. While these scholars serve as the foundation for theorization, they have largely zeroed in on the analysis of performance products themselves, with less specific focus on the role of the audience. The link to spectatorship comes from a couple of key texts from theatre and performance studies, including Andy Lavender's (2016) *Performance in the Twenty-First Century: Theatres of Engagement* and *Mapping Intermediality* (2010) by Sarah Bay-Cheng, Andy Lavender, and Robin Nelson. I take up the theoretical and practical foundations set by this generous assemblage of scholars and practitioners and extend their contributions to look at contemporary acts of spectatorship, with the primary focus being the experience of the individual spectator.

The approach also merges the above foundations with correlating theory from media studies, reception studies, games studies, sociology, and phenomenology to develop an interdisciplinary mode for spectatorship studies. This interdisciplinary focus allows one to explore the reciprocal relationship among the technologies people in deeply mediatized social systems use daily, the performance of everyday life in those systems, selfhood informed by those systems, embodied perception and interpretation, and the increasing prevalence of experiential forms of performance. The framework allows one to analyze spectatorship and performance where they intersect with(in) technological paradigms of mediatized sociality. Interactive technologies in the twenty-first century encourage a relational condition in society that makes more visible the interconnections between all elements in the mediatized social sphere. The mediatized condition informing this project helps to explain how twenty-first-century technologies and mediatization begin to influence the overall makeup of social systems to fit a specific form of relationality divorced from the human-centered epistemology of earlier unidirectional sender/receiver-based forms of media. This allows students, scholars, and practitioners of performative media a way to both analyze and plan for audience behavior across various forms.

As the mediatized spectator becomes the focus, the emphasis of analysis is less situated on a relationship between seeing and being seen, and instead on a coequal relationship between inputs and outputs of relational exchange. Individual technologies and the overarching technics that inform cultural and social milieus influence the shape, intensity, and agency of these inputs and outputs. Due to the current technics embedded within digitalization and datafication, mediatized spectators engage in interactive feedback loops with(in) both narrative and spectacle. This feedback informs modes of perception and the ways we process our individual and social realities. Previous modes of social communication were often explained through constructions of in-person face-to-face communication, and media simply transferred the information of that communication in one direction: From the media to the person. What has changed is how contemporary media platforms are increasingly more interactive and promote heightened levels of reactivity and relationality that people must further develop to perform effectively. In performance events, the trend toward relationality has slowly been building strength over the twentieth- and twenty-first century alongside the interactive technologies that have become more embedded in society. At the end of the twentieth century, this trend became an important area of scholarship and practical exploration.

With the English translation of Nicolas Bourriaud's *Relational Aesthetics* (2002) along with the publication of Jacques Rancière's *Emancipated Spectator* (2007:2009), performance and theatre scholarship on the nature of spectatorship has flourished in the early part of the twenty-first century. If the twentieth century was considered the century of the director, the twenty-first has arguably become the century of the spectator. As of the writing of this project, Rancière's book has a citation count of over 4,000, and Bourriaud's (2002) nearly 6,000 (Google Scholar). Alongside the rise in investigation into interactive and participatory modes of theatre and performance art, inquiry about spectatorship and audience studies has grown. The emphasis on much of the literature in theatre and performance studies has focused on the ways the spectator and their collective audience create meaning through modes of engagement, interactivity, or affective relationality. As Reason et al. (2022) explain, much of this has had to do with the "paradox of passivity" through which previous studies defined the audience as a body that does nothing physically yet is still crucially important (7). Reason et al. however highlight the study of the audience has a slight but important shift from the individual study of the spectator. As Ric Knowles (2014) points out in a special issue for *Theatre Journal*, recent scholarship "tended to write less about audiences and more about spectators or spectatorship" (xi). While the exact reason for this is difficult to pinpoint, I suspect it has to do with the fact that it is increasingly difficult to analyze and define an experience for an entire collective audience, whereas it is less difficult to do so with an individual spectator. The rise in what has come to be known as the experience economy (Pine and Gilmore 1999) has much to do with this as well.

Alongside the rise of the experience economy, developments in Human-Computer Interaction (HCI) that guide digital interaction became increasingly important. With communicative technologies becoming much more centered on the individual (user) versus a mass audience, the emphasis on individual experience became more important to understand (McCarthy and Wright 2004). In HCI, the emphasis is on how the human interacts with the technological tool, which creates the overall experience of tool use. The importance of HCI opened an entire new field of User Experience Design (UX), where the product was only one part of the Human/Interface circuit of experience. It is in this manner that an approach to technology requires a materialist phenomenological perspective, as the interaction with tech becomes the quotidian phenomenon impacting daily experience. This approach forms the basis for virtually all user experience design today. Following this approach, one can see how the logics, philosophies, and methodologies embedded within HCI and UX have seeped their way into the shaping and questioning of experiential performance practices in the last 20 years. The broad range of theatre and performance studies scholars focusing on experiential performance have undoubtedly been informed by these fields, whether they acknowledge them directly or not.

Within the realm of theatrical spectatorship studies, key texts that explore the various architectures covered in this project include Clair Bishop's (2012) *Artificial Hells*, Gareth White's (2013) *Audience Participation in Theatre,* Jen Harvie's (2013) *Fair Play*, Josephine Machon's (2013) *Immersive*

Theatres, Andy Lavender's (2016) *Performance in the 21ˢᵗ Century*, Adam Alston's (2016) *Beyond Immersive Theatre*, Rose Biggins (2017) *Immersive Theatre and Audience Experience*, Liam Jarvis' (2019) *Immersive Embodiment*, and James Frieze's (2016) edited collection *Reframing Immersive Theatre* to name just a few. Each builds off Bourriaud and Rancière in some capacity. Bourriaud's critique and theorization about the reception of relational art are grounded in the sociological theory of Pierre Bordieu, and Rancière's primary inquiry questions the political and aesthetic effects of relational and participatory performance to better understand the political potential of individual spectators. Both theorists outline the impact of such works, but neither fully considers the impact of contemporary mediatization and technogenesis as part of its rise. As Reason et al. (2022) explain, it was not until this past decade or two that study of audiences moved beyond a focus on either reception theory in the performing arts or the impacts of media and cultural studies on popular forms of media such as film and television (9–10). Coinciding alongside Rancière and Bourriaud, the focus of analysis on the performance of spectatorship and audiencing, however, began to reconfigure the relationship to be one of co-work between media and spectator. One might question whether their arguments highlighted a change occurring, or encouraged more questioning on the work, which in turn brought more visibility to what was already there. The fact that both were engaged in better understanding spectatorship from a socio-political perspective most likely helped spur heightened interest, as the performing arts have been reflexively on defense to the encroachment of technological media for the greater part of the twentieth century.

Bourriaud foretold the current era of techno-sociability through the digital as a coming fundamental shift in the ways of operating in the world, warning of "epistemological upheavals (concerning new perceptual structures), stemming from the appearance of technologies" (66). His writing is concerned with a shift toward relationality (social interaction) in the art world beginning in the early part of the twentieth century and culminating in the late 1990s. He lists the precursors of relational artwork as the projects of Dada, the Surrealists, the Situationists, and the Fluxus Movement. These movements are often considered the forebears of contemporary postdramatic and experiential performance. His work defines a new era of sociality through relational art with communicative social implications. The role of art was to model "possible universes" and to enact "ways of living and models of action within the existing real" (13). His *relational aesthetics* are underpinned by a "materialism of encounter" in which the "essence of humankind is purely trans-individual, made up of bonds that link individuals together in social forms which are invariably historical" (18). Relational aesthetics offers a way of understanding participatory and interactive forms of art, but also a changing world view and sense of sociality. The aesthetic objects documented and theorized, include all encounters within the "sphere of inter-human relations"

(28) which can become a frame for the production of art; as such, any relational act has its spectators who enter into a process of exchange. Bourriaud argues that the Art world—capital A—was responding to a new phase of social configuration based on participation, interactivity, and relationality, which in turn was predicated on a form of social politics (Freshwater 2009, 59–60). Much like mediatization, his relational form is concerned with states of becoming, change, and beingness in flux. These states require reciprocity from another agent or interactor to continue being and affecting. Spectators of these forms of art become part of the art itself through their individualized and communal agency. The relational form relies on the ability to interact which requires the spectator to become an agent in relation. Meaning emerges from this enactment of agency. The spectator performs as a "joint creator of the work" (Bourriaud, 99). Like many of the authors cited in this project, Bourriaud's theories are deeply influenced by the philosophy of Gilles Deleuze and Felix Guattari, as an extension of the poststructuralist trend toward breaking down definitive bounds of meaning and existence. While Bourriaud's contribution to the field of spectatorship studies in theatre and performance is significant, the importance of his work has largely been overshadowed by Rancière's.

After the publication of Rancière 2007 speech, the annual output of published scholarship on spectatorship nearly doubled.[3] His reclamation of the spectator as an active participant with political capacity ushered in an emphasis on audience and reception studies in theatre, film, and performance. Rancière's theory calls into question the political efficacy of all theatrical and art experiments aimed at energizing a "passive" audience into new modes of political agency. Emancipation starts "when we realize that looking is also an action that confirms or modifies that distribution, and that 'interpreting the world' is already a means of transforming it, of reconfiguring it" (2007, 278, quotations in original). He argues that because the ontology of spectating is already an active function, spectatorial participation in theatrical performance is unnecessary to strive for. Part of his argument explains how the binary of watcher and performer was constructed to establish modernist models more firmly established to advance consumerism and commodification. From this basis, he argues that for emancipation to occur—freeing the spectator from the grips of the commodity fetish— there is a necessity for the performance medium to operate as an equal contributor in knowledge creation with the spectator. Thinking in this manner allows knowledge and meaning-making to emerge as a form of co-work, where both the spectator and the spectacle have emancipatory potential. The spectator becomes free from the bonds of thinking about their performance of spectatorship as a passive consumption of art. I tend to agree with this argument. Following both recent discussions of neurological analysis and spectatorship (Cook 2008; Damasio 2005, 2010; McConachie 2015; McCutcheon and Sellers-Young 2013) as well as theories of media readership (Jenkins 2006; Wilson 2008), the process of watching is never passive as it is always relational. Watching always involves some form of interpretation, either at the conscious or unconscious level. Spectating also engages a watcher's perceptual apparatus in processes that can allow them to evaluate the event critically and sensually.

Experiential spectatorship, however, moves beyond the primacy of watching and instead focuses on concepts of *being-in-the-event* and *interacting-with-the-event*; concepts that are more easily understood when put into conversation with twenty-first century media and technologies. Rancière might be so popular among theatre and performance scholars because his primary objects of analysis were the theories of Brecht and Artaud. However, they were active in the mid-twentieth century, well before the advent of interactive media such as the personal computer, the internet, the smartphone, and social media. Twenty-first century spectators are already active and participatory in the ways they interpret and consume information as part of digital culture and mediatized society. With each of the communications technologies mentioned above, human sociality in deeply mediatized cultures has increasingly become more involved and interactive in the stories digested, not simply through spectating, but more so through the personal sculpting of the production, dissemination, and consumption of both narrative and event (Hepp 2013; Jenkins, Ford, and Green 2013). Due to this conditioning of contemporary spectators through technogenesis, the binary of passive and active is less helpful when thinking about acts of experiential spectatorship. Instead, it is more helpful to think in terms of relational exchange.

Recent scholarship expanding into relationality includes Lavender's (2016) *Performance in the 21st Century*. Though he delves deeply into the arguments of Rancière, he expands his scope outward to the relationship of spectators and their societal frames. Lavender describes a shift in societal and cultural milieus in the early part of the twenty-first century, leading to what he names "theatres of engagement." These theatres are prompted by a shift from a "society of the spectacle to a society of involved spectaction" (29–30). This quote harkens back to Guy Debord's (1995 [1967]) attack on the societal impacts of capitalistic media modalities of television in the mid-twentieth century that had lulled people into a consumerist passivity. For Lavender, contemporary spectators no longer simply watch though; they interact and engage with their social surroundings. Theatrical performance at the end of the twentieth century became "something other than an encounter between actors, or between actor and audience," it was evolving into a form where the "separation between the space of the performance and that of spectatorship" was quickly closing (9). Similarities exist between Lavender's engaged spectators and the mediatized model proposed in this project. The connections come from a necessity for understanding constant interaction with and feedback from the spectacle that is embedded with(in) postdigital mediatized life. This includes an evolution toward relationality and engagement that has encouraged "a sharpened enjoyment of co-presence, corporeality and embodied sensation" (15). In Debord's mid-twentieth century paradigm of spectatorship, the spectator acts as a receiver of information, primarily through sight and sound. The construction of a mediatized spectator asks one to think of spectatorship as a fully embodied practice undertaken through their mediatized perceptual apparatus. This apparatus is uniquely linked to the fluctuations in the technological environment and the technics that construct that environment. As Lavender notes, "we experience culture differently because we do so with our minds and expectations adjusted to the speeds and shapes, flows, and frames of the expressive apparatus with which we live" (19). Through technogenesis, contemporary media and technologies condition individual subjectivity and impact the expectations one has with the way they interact with performance and narrative.

This project moves past the models offered by Rancière and Debord to enter a landscape for considering interaction and engagement as prevailing modes of contemporary spectatorship based on feedback loops. Literature on the relationship between media/technology and performance often includes logics of feedback between technology and human beings. Feedback refers to the disruption of a signal based on duplicated inputs from an original source. We colloquially use the term to refer to the noise (signal) delivered from a speaker, which is picked up by an electronic mic that then feeds that same signal back to the speaker via amplification. When the original sound feeds back to the speaker, it becomes distorted as unintelligible noise. A feedback loop, then, is the constant cycle of inputs and outputs that creates change. Erika Fischer-Lichte (2008) discusses the autopoietic feedback loop that exists in (and defines) performance in her work, *The Transformative Power of Performance*. She explains that a performative gesture/act from a performer serves as an output signal, which a spectator reads/interprets and then returns to the actor. This then affects how they both continue to perform in some manner. There is a mutual interaction between the two and hence a feedback loop. Michael Darroch (2010) discusses this process from the perspective of the performer as one where the action of the audience initiates the loop. He argues that with the introduction of sophisticated stage lighting and the engulfment of the audience in darkness, the feedback loop created by visual spectatorship was interrupted. No longer could the actors see the audience, and likewise, the audience could no longer see their fellow spectators (186). This was the model Brecht and Artaud were responding to. The trajectory toward separating the audience from the spectacle in theatre was a long process that began emerging over the course of the seventeenth and eighteenth centuries. This process of change created a dynamic where the spectacle on "stage" became the central agent of analysis and the spectator was considered a secondary component. This remains true for much of the 20th century.[4]

From a technological perspective the concept of the autopoietic feedback loop was introduced during the first wave of cybernetics theory by Norbert Wiener (1948) and was used to define the reflexivity inherent in the dual-directional flow of information that makes up any body inside a system. The human body, as a perceptual apparatus, exists as part of the information system made up of the environment (ecological, cultural, material, techno-logical) it resides in. Through feedback loops, a self, operating under a medi-atized mode, has the potential to cooperatively create itself through a process of autopoiesis which refers to self-formation and self-regulation based on environmental input. For a mediatized spectator, the feedback loop is inher-ent in their subjective conditioning due to the way information travels not only between the spectacle (environment) and the spectator, but *through* and *within* both, creating not only their exterior view but also their inner makeup or being.

Like feedback, the terminology of *exchange* is fundamental to my framework for spectatorship as it helps to better explain interconnected relationships. Embedded within discussions of mediatization is the idea that people—and specifically unmediated person-to-person forms of communication—need to be decentralized as sole agents of exchange in me-diatized models of social life. Utilizing this theorization offers a framework for analyzing the relationship between audiences and events that focuses on spectatorship that is interactive, relational, and co-directional. Experiential spectatorship then involves multiple agencies: That of the spectator (its per-ceptual apparatus), the architecture the spectator performs with(in), as well as the technological processes and cultural products creating technics, that inform modes of spectatorial perception with/in/through performance. In previous theories of communication and agency, people were understood as the primary objects/subjects. In a mediatized model, agency is distrib-uted across the entire ecology in which human beings, technologies, media, and social processes operate. Technologies are a crucial part of this ecology and have agency that is inseparable from the human agents involved. This study offers a more thorough understanding of how structures of twenty-first century performance—and subsequently the performance of spectatorship—are related to changes in perception as informed by technogenesis and deep mediatization.

The ultimate goal of the project is to offer an analytical viewpoint for better understanding the relationship between individuals and the social constructs they create; technologies and the technical paradigms they influence; and a predilection toward experiential spectatorship in the twenty-first century. I begin by explaining the relationship of the individual to mediatized culture and how this relationship brings about changes in society. I then move on to laying a basis for understanding how contemporary technogenesis leads to multiple impacts on individual perception based on the influence of technics. I then discuss how a mediatized spectator operates through the four various architectures of exchange as distinct but overlapping sets of operations and aesthetics for experience. The viewpoint offered can be used as a lens for analyzing how contemporary spectatorship is connected to societal changes spurred on by technological interfaces in contemporary cultures. By understanding these changes, the analysis of spectatorial acts with(in) experiential performance events, the events themselves, and hopefully the creation of these events allows new possibilities geared toward addressing the needs of a changing demographic of audiences.

Architectures of Exchange

A mediatized spectator is one who is constantly in flux and navigating multiple relational actions informed by the specific constraints of the architectural system with(in) which it operates. This mode of spectatorship relies on an embodied phenomenological subjectivity. A mediatized spectator could be thought of as an interactor, a relationist, or as what Robin Nelson (2010) calls an experiencer. For Nelson, the spectator as experiencer,

> suggests a more immersive engagement in which the principles of composition of the piece create an environment designed to elicit a broadly visceral, sensual encounter, as distinct from conventional theatrical, concert or art gallery architectures which are constructed to draw primarily upon one of the sense organs – eyes (spectator) or ears (audience). (45)

In each chapter, I further break down this interactive, relational, and experiencing spectator into distinct sub-spectators based on the *architecture of exchange* they perform with(in). These include the *immersant*, the *participant*, the *player*, and the *(co)author*. In each, an emphasis on perception, agency, exchange, bodily affectivity, and interconnectedness between multiple senses, critical faculties, and technological objects occurs. This requires further exploration of each architecture and how they relate to and inform how a mediatized spectator performs. The experiential model of spectatorship also relies in part on an ecological perspective, as it is not only informed by technology but also formed by the changes in cultural politics that interaction with technologies as social actors allows. As such, I've included examples in each chapter that have a connection to social, ethical, cultural, and/or political change.

In an earlier connected project (Lewis and Bartley 2022)[5], I included a brief definition of the individual architectures without deeply defining their relationship to contemporary technologies and technics. The goal there was to introduce the terminology in a way that allowed new forms of pedagogy to develop for educators training contemporary theatre students for modes of experiential performance making. In that project I could only scratch the surface however. The individual architectures are offered here to both explain how they have become increasingly prevalent over the past forty-odd years and to frame the experiences of contemporary spectators during that same time period. They are the most readily visible examples of experiential format fitting into the interlocking domains of contemporary media. I utilize the four architectures of exchange in the following manner (Figure 0.1).

Immersion allows a spectator to be thrown into a fictive world, giving them the impression of being a member of that world and allowing heightened levels of *perceived agency* based on a sense of beingness with(in) that world. For example, in the highly influential immersive theatre production *Sleep No More*, an immersant submerges themselves in a free-flowing narrative experience by navigating scenographic space and feels a sense of agency through their ability to navigate at will. The degree of perceived agency is a crucial element to the architecture and experience of immersion. Agency arises as an affective byproduct of the relationship between a spectator's perceptual apparatus, scenic framing, the narrative, and the aesthetic components of the immersive media. My focus on immersion is its relationship to the technic of **virtuality** which has informed how people understand the border between the virtual and the actual. This border is often crossed and blended through various technologies over time but became increasingly influential during the rise of the home computer generation. Through the home computer and then the extension of time and space through the internet, virtuality began to imprint on society a condition through which individual subjectivity started to become untethered to the physical surroundings of the actual. This condition would then lead both artists and spectators to seek out experiences using logics defined by the possibilities of the digital.

Participation operates by allowing modification of the event that can be witnessed or documented by the participant or others. Participants have agency to interact in a material manner, creating moments of rupture in the narrative telling or world. Though this rupture does not always dictate the final outcomes of that telling or world, it does create a tangible impact. This capacity to materially participate has always been part of spectatorial acts. One might argue that it is the basis for all performance. However, with the rise in mechanization and electrification, theatrical practice created an artificial divide between the audience and the medium, delimiting the spectator's importance to the event's totality. Instead, the product was placed in direct relation—most likely in front of the audience's eyes or ears—with the expectation that the product would transmit the information to the audience with little regard for a return of information. This then became the dominant mode of audience behavior as mechanical and then electrified means of media and performance evolved. With the advent of the internet, relational interactivity began to slowly replace the dominance of these reception-based media and formats. As the internet evolved, it became increasingly participatory, allowing the user multiple ways of contributing and collaborating with both the medium and others consuming media. This allowed the technic of **networked collaboration** to arise as another influence shaping sociocultural perspectives on reality. Both immersion and participation introduced

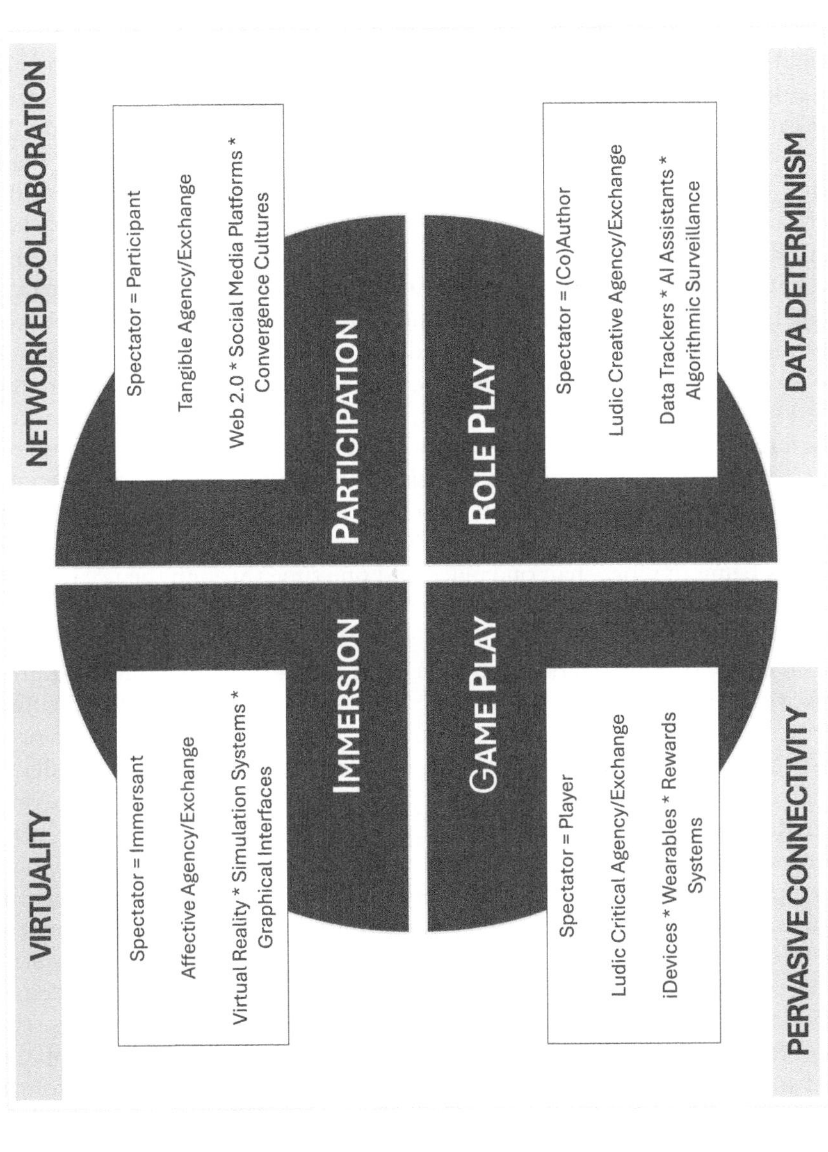

Figure 0.1 Taxonomy of Experiential Exchange.

significant breaks from the dominance of twentieth-century media and performance. Like immersion, participation is often used as an umbrella term but will be discussed as only one architecture of the unified framework of experiential spectatorship.

In *Game Play*, a mediatized spectator becomes an increasingly active member of the world they participate and/or are immersed with(in) due to a formal design structure based on rules and objectives. These rules dictate both story and mechanics for engagement as well as the way a spectator creates meaning. Rules frame the way a player can reach specific goals. In many instances, these rules are what define games as specific forms of ludic activity. A gamified experience allows spectators the ability to create meaning at the threshold between the game-world and the real-world which has an unstable overlap. As the logics of the participatory web became commonplace within digital societies, corporations quickly learned how to harness the potential of participation by instilling gamified operations and structures on the increasing number of connected digital devices. This led to the technic of *pervasive connectivity* through which smart mobile devices serve as the primary conduits. Through this technic, the digital and actual become intricately intertwined in a way that makes it nearly impossible to differentiate the two. These devices also fuse the logics of both immersion and participation into an always-on and always-connected tool for human/technology inter-connectivity. Through this fusing, people become habituated to a sense of always being in more than one place and one time through the use of their devices. Games and gamification serve as the central architecture that keeps users from detaching from their devices, creating a symbiotic relationship between the two. This relationship then promotes a need for continued processes of game play. In games and game play, spectators often gain potential for critical agency and self-reflection. Game play can be harnessed for critical retrospection, however, when used purposefully, negating some of the detrimental effects of the tool use.

Via *Role Play*, a spectator engages in a constant negotiation with reality through imaginative authorship and unlimited potentiality to define their understanding of the self. Role play works as a ludic form of self-authorship for the mediatized spectator. While role play has been theorized as a central element of all social life, over the past decade the technic of *data determinism* has established a powerful social force with the potential to redefine individual selfhood. Through the rise of algorithmic tracking processes, the corporations and governmental actors who harnessed the capacity of the digital sphere gained an incredible amount of power to dictate the outcomes of the "game of life" digital users are embedded in. This data-based mode of play is also based on a system of invisible surveillance. This is referred to as dataveillance and is a process through which individuals are defined by the choices and actions they make in the digital sphere. As a mode of experiential spectatorship, role play does, however, offer a tool for both understanding and combatting the power of data determinism. Under data determinism, all reality has the potential to become a constant mode of ludic techno-performativity with(in) ecologies of postdigital relationality. Through role play, the spectator performing as a co-author gains the agency to dictate, reconfigure, and reassign an understanding of their identity and the experience they undertake via experiential exchange.

The Chapters of the Book

When the above architectures are put into relationship and conversation with each other, they allow for a conceptual framework for experiential spectatorship. To construct this framework, the project begins with mediatization and its relationship to perception and sociality. Chapter 1, **Mediatization, Perception, and Experience**, begins with section **1.0 "Foundations of the Mediatized Spectator"** that presents the reader with a condensed understanding of the process of mediatization and the impact of media as a communicative link in contemporary society via Couldry and Hepp (2017). Following their concept of mediatized social environments, media of various modes cannot decouple from the communication process. Whereas previous models of social constructions relied on a model of personal human-to-human communication (Berger and Luckmann 1967), those models are less useful due to the pervasive presence of digital media as an intermediary between human beings and the entirety of the world. Couldry and Hepp are quick to qualify that this level of pervasiveness is not uniform across the planet, but by describing waves of mediatization and technological change, they persuasively argue that we have entered an era of human sociality that can no longer operate without these media. Their model for sociality informs the spine of the entire project and therefore is revisited in part throughout the remaining chapters.

The chapter further refines the argument about technology and human perception through an introduction to technogenesis and technics. The process of technogenesis is the formal theoretical device applied throughout the project to better understand human evolution based on relationships with technology via technics. Covering technogenesis bridges the gap in understanding spectatorship before and after the advent of deep mediatization. To explain technogenesis, I describe changes in perception informed by adaptations in the way people read from printed text to hypertext. This description is informed by N. Katherine Hayles' (2012) book *How We Think,* which connects to her other work on technoculture (1997, 2005), digitally augmented cognition (2014, 2017), and technologically informed posthumanism (1999). Hayles corpus of scholarship considers the relationship between people and technology as a way of redefining the way humanists approach literary analysis. The rest of the chapter then introduces and defines the concept of the mediatized spectator's perceptual apparatus, which operates as a fluid and in-flux amalgamation of the individual's body, mind, and technological environment. This apparatus is a "posthuman" machinelike construct that works alongside technogenesis through embodied perception. This interacting and interfacing machine appears throughout the project as the mediatized spectator's mediator between technologies and the world. This construct is crucial in describing the materialist phenomenological approach to mediatized conceptions of reality formation through spectatorship. The sub-section also includes a nod to this techno-materialist approach found in various literature from intermediality studies. This review points to a similar viewpoint on the mediatized spectator but through the lens of the posthuman which is lurking almost hidden in the background of many of the theories engaged through the rest of the project.

The remaining chapters are broken down based on the individual Architectures of Exchange, with each chapter split into two primary *hinged* sections using a 2.0/2.1 naming structure. This titling reflects the relationship to digital cultures where theory and application meet through experience. Each chapter offers the theoretical and historical on one side and a set of readings of spectator experience referred to as Instances on the other. The effect here is to combine two perspectives into one. This allows a dynamic where the reader has the agency to read the Instances first and then go back to focus on the broader framing if they choose. The first section of each chapter offers a supplementary set of theories that sit underneath the broad umbrella of mediatization and guided by technological affects discussed as technics. A short historical introduction to the specific technologies that inform these technics then moves into the individual architecture. The layered approach from theory to specific technologies and technics leads to a more defined lens for each architecture covered at the end of the section. The goal here is to set up a lens for how the mediatized spectator operates through these specific technics and then discuss the modes of exchange that naturally fit with each mode of operation. The goal is not to promote an argument that each technic is solely responsible for the individual architecture, but instead, to use it as a narrowed perspective for analyzing spectator experience with(in) each architecture. The way the four architecture chapters are laid out is based on a historical trajectory through the technologies and technics, but the reader might choose to jump straight into whichever architecture most suits their particular needs. I could see some interest in starting with Chapter 5 and moving backwards or in jumping around by reading all the theoretical framings and then the Instance sections. Each Instance is framed by a specific architecture, but as there are multiple overlaps between how one might analyze a specific Instance, one could choose to read it against the architecture offered if they so choose. As agency is a crucial concern in each, I welcome the reader to engage with the architecture chapters in whatever manner suits them best.

Chapter 2, *The Architecture of Immersion,* begins with section 2.0 "**Virtuality, Simulation, and Technologies of Immersion**" where the reader is asked to consider what they understand as real. This question is at the heart of the technic of *virtuality* which has been impacting the relationship between immersion and performance since the musings on Plato's cave. The chapter focuses on how virtuality became increasingly powerful and important with the advent of the home computer age, where virtual worlds started to become accessible for everyday consumers and evolved into a growing interest in the collective consciousness through the fascination with virtual reality technologies, cyberspace narratives, and adoption of internet protocols. Theoretical and historical conceptions of virtuality and simulation by authors such as Brian Massumi (2002), N Katherine Hayles (1997, 1999), Jay David Bolter and Richard Grusin (1999), Janet Murray (2017), Pierre Lévy (2001), and Jean Baudrillard (1995a, 1995b, 2014) are offered to develop a groundwork for how technologies of virtuality operate through immersion as part of the technogenetic relationship between spectators and virtual reality games and immersive events. This leads to the Architecture of Immersion and its relationship to the act of spectatorship between two different forms of reality: One that exists in a material world, often called actuality, and the other a realm of the not quite real or virtual that becomes actual through its potentiality. This potentiality is analyzed through its relationship to immersion as a process connecting spectators to virtuality through bodily affect. The primary mode of exchange that emerges in immersion is sensual affectivity, a mode of information transfer and communication that relies on the feeling body of the spectator. The first hinge section, **2.1 "The Feeling Spectator and the Affect Economy of Immersion"** serves as the initial Instance chapter including multiple experiences of immersion in experiential media. It begins with a retelling of my personal experience with the Virtual Reality (VR) event *Ghostbuster's Dimension* (2016), housed within the interactive Ghostbusters exhibit at Madame Tussauds wax museum in New York City. This is followed by another first-person account of the VR experience playing the PlayStation shooter game *Farpoint* (2017), which further defines the relationship between immersion, the spectator's body, and affect. Next is an analytical breakdown of the theatrical performance *The Encounter* (2016) by the London-based performance company Complicité. The production employs immersive audio to augment the experiential quality of a proscenium style one-person play. The analysis explains how the production uses binaural audio as a tool for how immersion operates with and on a mediatized spectator's perceptual apparatus. As questions about spectatorship often rely on conceptions of visuality, this case study offers a different understanding of virtuality and immersion accessed through other aspects of a person's feeling body. Sensual affectivity is then discussed as the primary mode of agency through an analysis of an experiment with computer-assisted virtual immersion in Punchdrunk's 2012 collaboration with MIT on the show *Sleep No More*. While *Sleep No More* has been widely discussed elsewhere, this analysis sets up a distinction between affective and tangible exchange. I finish the chapter with a brief argument about the pervasive presence of immersion in recent scholarship and its relationship to individualistic consumer culture, putting into conversation Adam Alston's (2016) critique of the experience economy.

Chapter 3, *The Architecture of Participation*, begins with section 3.0 "The Promise of Web 2.0: Participation, Convergence, and Collaborative Technologies" where the technic of **networked collaboration** is covered to explain how Web 2.0, and specifically digital platforms for social media have promoted and supported forms of experiential spectatorship based on choice making and material change. The chapter begins by asking the reader to imagine the possibilities presented to them in a paradigm where individual choice is predicated on the operations of a social sphere maintained by the network logics of the internet. Participation is approached from an active register in that a mediatized spectator gains a form of tangible agency rather than the affective agency described in the previous section. Henry Jenkins (2006) theories on participatory and convergence cultures are put into conversation with Manuel Castells (2004) network society. This offers a way of thinking about the experience of participation in performance as something intricately interconnected with the logics of Web 2.0. This wide-ranging logic, built into the collaborative nature of the internet, is then explored through the various social media platforms that came to dominate digital communication in the first decade of the twenty-first century. These media are based on tangible and material forms of participation coming directly from the user. This is the basis of the architecture of participation, where the exchange is tangible, and the spectator has the potential to make a material impact in/ on the event. I argue that the connection people have become accustomed to through the Web 2.0 tools has encouraged a form of democratic agency expressed in participatory performance. The connections between participatory culture and the technic of networked collaboration are then explored in a set of participatory media. Section **3.1 "The Democratic Spectator: Choice Making and Political Exchange in Participatory Performance"** contains a set of Instances that show the ways the spectator experience is directly related to the choices they make. A brief playthrough description and then analysis of choices offered in *Detroit: Become Human* (2018) for the PS4 game system sets up the ways choices impact story in a choose-your-own-adventure style digital narrative. As the narrative focuses on the ethical relationship between human and non-human entities, the choices made become political and have the capacity for shaping democratic ideals for the participant. The next Instance shifts over to a live participatory theatrical production where the audience uses online devices to collectively make decisions that shape a *Dungeons and Dragons* style fantasy adventure unfolding before their eyes. To better understand how performance makers might craft such possibilities for choice making as a collective audience experience, an edited conversation with the makers of *The Twenty-Sided Tavern* presents opinions on audience agency, ideal audience experience, and the relationship between participatory culture and contemporary audiences. The final Instance centers on spectatorial participation developed out of the Occupy protest movement of 2011. As social media platforms were reaching their peak potential as tools for social change, The Civilians developed a participatory work titled *Occupy*

Your Mind (2011) intended to propel the activist energy of the movement forward in time. By allowing future audiences the experience of (re)living and re(performing) the verbatim narratives of Occupiers using social media tools, they gain a direct and tangible entry point into the democratic ideals of participatory activism. The chapter ends by questioning the capacity a participating spectator has to engage with ethical and communal concerns as technologies evolve. What agency do they have to create a form of exchange with social and political energy when the forces of capitalistic production subsume the power of participatory culture?

Chapter 4, *The Architecture of Game Play*, starts with section 4.0 **"Pervasive Connections: Smart Devices, Locative Media, and the Gamification of Reality"** and returns us to the binary of the virtual and the actual presented in Chapter 2 by asking the reader to imagine the ways in which mobile devices create a hybrid reality where the two meet. This hybridity comes about through the ways smart devices become technological extensions of one's sense of being-in the-world that is reinforced through the processes of gamification. The technic of ***pervasive connectivity*** is offered to show how mobile technologies augment our perception of the Real, which in turn is read as a mode of spectatorship through the architecture. *Game Play* operates as a process and structure that can be best understood through the logics of mobile technologies, which are discussed as iDevices. These devices are highlighted as technological appendages whose technogenetic relationship to one's perceptual apparatus allows a dynamic where a mediatized spectator enters into a game-like, liminal state of techno-embodied perception and operation with the world. Mediatized spectators augmented by mobile media gain access to place, space, and time in a manner that transcends conventional modes of watching and even interacting to develop a perceptual function that is more akin to being forever suspended in bounded, yet liminal play. The primary theoretical spine of this chapter follows Jason Farman's (2012, 2021) theorization of mobile embodiment and the concept of liminal hybridity from Adriana de Souza e Silva (2009). Pervasive connectivity is explained as an always-on and always-connected affective process that brings about phenomenological changes to the perceptual apparatus that allows embodied experience to transcend the confines of the actual. This technic is then placed historically through an unpacking of the incorporation of mobile devices in society. As a way of returning to the early conversation regarding generational shifts, a brief section on the relationship between Generation Z—often called the iGen—and media use segues into theorization regarding play and specifically modes of game play based on formal structures and design that lead the player toward defined objectives. *Game Play* combines the affective and tangible registers of experience in *Immersion* and *Participation* with the possibility of a meta-agency of critically reflexive examination with respect to a structural understanding of the game-world. In the examples of game play and gamification covered, the spectator's experience of exchange is based on becoming a critically activated member of the worlds navigated, due to an established set of ludic rules. Section **4.1 "The Playing Spectator: iPerformance and Ludic Criticality"** covers three Instances of game play to show how the mediatized spectator's experience is augmented when using iDevices. Each example also offers a twist on the way one sees the world around them through an ethical lens, which is described as ludic criticality. The first Instance is a personal encounter with the augmented reality game *Pokémon GO* (2016). This reflection allows the reader a better understanding of how iDevices interrupt conventional understandings of space and place when their locative capacities are used in gameplay. When playing the game in the historic Santa Fe town square, the histories of the place are uncovered to show the ways the virtual and actual overlap, offering critical understandings of the player's place in the world. The next Instance is the UK company Coney's alternate reality game *Adventure One* (2015). The performance uses mobile devices as a figurative prosthesis for playing spectators, bringing them into the narrative while also serving as a tool to frame a critique of location-based politics and global economics. The final Instance is a brief play-through of the smartphone app-based game *Phone Story* (2011) that puts into relation the players gamified actions and the operations of the iDevice. The chapter ends by considering the capacity of meaningful play and the ethical questions that arise when the player is placed within the global ecosystem the iDevices exist within.

Chapter 5, *The Architecture of Role Play,* begins with section 5.0 **"Algorithms, Avatars, and Affective Computing: Datafication and Social Feedback Loops"** and moves forward to cover datafication as the final wave of mediatization offered by Couldry and Hepp. Datafication exists as a social force where all media is filtered through the surveillance, capture, computation, and redeployment of digital data. The reader is asked to examine how well they understand their own self. Is that understanding more or less real than the data doubles created by algorithmic technologies? The technic of *data determinism* explains the multiple invisible forces of data-based processes in contemporary life. By pairing the work of Couldry and Hepp (2017), Hayles (2014, 2017), Mark Hansen (2015), and John Cheney-Lippold (2017), I explain how algorithms and artificial intelligence are defining a new era of technogenesis that highlights the power of role-play as a performative mode of spectatorship and life. Through data determinism, a mediatized spectator's sense of self is put into constant tension with the avatars and data-based constructions they both author in digital spaces and by data tracking algorithms. Through data determinism, spectatorial acts with(in) the architecture are read as performative experiences of ludic creativity. This leads to strategies in and of performance that highlight role-play embedded in digital and postdigital culture. Section **5.1 "The Spectator as Character: (Co)Authoring Identity Through Role Play"** focuses on three Instances of role play in online and offline media. The first is a part auto-ethnographic, part critical analysis of playing antihero characters in Rockstar Games' projects for the PlayStation platform. First is the notorious character Trevor in the widely successful smash and grab heist video game *Grand Theft Auto V*. Trevor represents the most vile and socially unacceptable character types in popular media culture, yet he is one of the protagonists. Playing Trevor is contrasted with the dynamics of the morality/honor system while playing Arthur Morgan in *Red Dead Redemption 2*. By examining the ways playing these characters are discussed in online forums and by my own play, one begins to understand the possible advantages of playing against one's own personality in digital domains. This is precisely where role play becomes socially relevant and applicable to new understandings of techno-performativity. This mode of exchange adopts aspects of immersion, participation, and game play while introducing ludic creative operations through which spectators use their imagination to actualize potentialities concerning the creation of multiple selves. The next Instance takes up the previous tools to examine ways one might engage in interpersonal dialogue while in the skin of another identity. In Blast Theory/Hydrocracker's covert surveillance experience *Operation Black Antler* (2016–2019), the spectator role plays the persona of an extremist in order to stop a potential activist attack. In this experience, they learn how one creates and perpetuates stereotypes as legitimate identities and how these stereotypes are reinforced by algorithmic technology. The final Instance is a reading of the television series *Mr. Robot* where the spectator becomes a "friend" to the protagonist Elliot, and by proxy part of the narrative, as Elliott confronts both his inner psyche and the forces of datafication in capitalist society. The chapter ends by re-focusing on techno-performativity by reading these experiences through data determinism. Through techno-performativity, a mediatized spectator learns how role-play is encountered and engaged in algorithmic culture. They then might harness the liminal position of the self as one full of potential to subvert the powers of algorithmic control.

The final **Afterword** puts all the architectures into conversation with each other to explain how each can be used as an analytical tool that teases apart specific modes of spectatorial exchange within deeply mediatized cultures. By doing so, one can clearly articulate the difference between individual experiences for the spectator. The final words also point toward the ongoing development of new technological tools, specifically advanced artificial intelligence agents that will undoubtedly lead to a new technical paradigm in the coming decade. As technology is continually evolving at an increasingly rapid pace, we must do our best to understand its effects and affects on the individual and society to create the best analytical tools for the ongoing evolutions in experiential performance. The project offers the reader a way of understanding the ongoing and evolving relationship between mediatization and acts of spectating in relational and experiential formats. By questioning, exploring, and analyzing the relationship of the spectator to the mediatized social sphere, readers gain a critical set of tools to rethink the relationships between experiential performance, society, and postdigital cultures. Thinking in a technological and critical manner helps to rethink experiential spectatorship as a relational process based on the logics of technogenesis. Using the individual architectures as signposts for the various substructures of experiential spectatorship allows a reader the building blocks to assemble and reassemble networks of technological interaction as modes of performance. Doing so may allow us to find new ways of connecting with future generations of spectators through our performance events.

Notes

1 My partner informally tested this response on four-hundred hundred twenty-five third through sixth graders during Spring of 2018. The response rate of those who made a similar gesture to the child in the anecdote was 35 percent for the oldest group and over 65 percent for the youngest group.
2 This mode develops in advanced technogenetic paradigms of deep mediatization and is not uniformly established across cultural boundaries.
3 Worldcat.org keywords search for Spectatorship, Spectator, and Audience with the qualifiers performing arts or art/architecture.
4 Audiences become a key point of interest for other media such as television, radio, and film beginning in the 1930s. See Sedgman 2016, 2018, 2019 and Freshwater 2021 for brief historical analyses of this shift.
5 The basis of this taxonomy and its relationship to technogenesis first appeared in "Performing Posthuman Spectatorship" (Lewis 2017), where I utilized the frame of critical posthumanism as an analogue to deep mediatization. The article serves as the core foundation of this book and has been reworked and reprinted in later chapters with permission of the editors of *Performance Research Journal*.

Bibliography

Alston, Adam. 2016. *Beyond Immersive Theatre: Aesthetics, Politics and Productive Participation*. Palgrave.
Auslander, Phillip. 2021. *Liveness: Performance in a Mediatized Culture*. 3rd ed. Routledge.
Barker, Martin. 2006. "I Have Seen the Future and It Is Not Here Yet …; or, On Being Ambitious for Audience Research." *The Communication Review* 9, no 2: 123–141.
Baudrillard, Jean. 1995a[1981]. *Simulacra and Simulation*. Translated by Shelia Glaser. University of Michigan Press.
Baudrillard, Jean. 1995b. *The Gulf War Did Not Take Place*. Translated by Paul Patton. Indiana University Press.
Baudrillard, Jean. 2014. *Screened Out*. Translated by Chris Turner. Verso.
Bay-Cheng, Sarah. 2007. "Theatre Squared: Theatre History in the Age of Media." *Theatre Topics* 17, no 1: 37–50.
Bay-Cheng, Sarah. 2012. "Theatre is Media: Some Principles for a Digital Historiography of Performance." *Theater* 42, no 2: 27–41.
Bay-Cheng, Sarah. 2014. "'When This You See': The (Anti) Radical Time of Mobile Self-Surveillance." *Performance Research: A Journal of the Performing Arts* 19, no 3: 48–55.
Bay-Cheng, Sarah. 2016. "Postmedia Performance." *Contemporary Theatre Review: Interventions*. http://www.contemporarytheatrereview.org/2016/postmedia-performance/
Bay-Cheng, Sarah. 2017. "Pixelated Memories: Performance, Media, and Digital Technology." *Contemporary Theatre Review* 27, no 3: 324–339.

Bay-Cheng, Sarah. 2023. "Digital Performance and Its Discontents (or, Problems of Presence in Pandemic Performance)." *Theatre Research International* 48, no 1: 9–23.

Bay-Cheng, Sarah, Chiel Kattenbelt, Andy Lavender, and Robin Nelson. 2010. *Mapping Intermediality in Performance*. Amsterdam University Press.

Bay-Cheng, Sarah, Jennifer Parker-Starbuck, and David Z. Saltz. 2015. *Performance and Media: Taxonomies for a Changing Field*. University of Michigan Press.

Benford, Steve, and Gabriella Giannachi. 2011. *Performing Mixed Reality*. MIT Press.

Bennett, Susan. 1997. *Theatre Audiences: A Theory of Production and Reception*. Routledge.

Berger, Peter L, and Thomas Luckmann. 1967. *The Social Construction of Reality: A Treatise in the Sociology of Knowledge*. Anchor Books.

Biggin, Rose. 2017. *Immersive Theatre and Audience Experience: Space, Game and Story in the Work of Punchdrunk*. Palgrave.

Birringer, Johannes. 1998. *Media & Performance: Along the Border*. Johns Hopkins University Press.

Birringer, Johannes. 2000. *Performance on the Edge*. The Athlone Press.

Birringer, Johannes. 2006. "Digital Performance." *Performance Research* 11, no 3: 42–45.

Birringer, Johannes. 2008. *Performance, Technology & Science*. PAJ Publications.

Bishop, Claire. 2012. *Artificial Hells: Participatory Art and the Politics of Spectatorship*. Verso.

Bolter, Jay David, and Richard Grusin. 1999. *Remediation: Understanding New Media*. MIT Press.

Bourriaud, Nicolas. 2002. *Relational Aesthetics*. Translated by Simon Pleasance, Fronza Woods, and Mathieu Copeland. Les presses du reel.

Broadhurst, Susan. 2007. *Digital Practices: Aesthetic and Neuroaesthetic Approaches to Performance and Technology*. Palgrave.

Broadhurst, Susan, and Josephine Machon. 2006. *Performance and Technology: Practices of Virtual Embodiment and Interactivity*. Palgrave.

Castells, Manuel. 2000. *Rise of the Network Society*. 2nd ed. Blackwell.

Castells, Manuel. 2004. "Informationalism, Networks, and the Network Society: A Theoretical Blueprint." In *The Network Society: A Cross Cultural Perspective*, edited by Manuel Castells. Edward Elgar.

Causey, Matthew. 2006. *Theatre and Performance in Digital Culture: From Simulation to Embeddedness*. Routledge.

Causey, Matthew. 2016. "Postdigital Performance." *Theatre Journal* 68, no 3: 427–41.

Cheney-Lippold, John. 2017. *We Are Data: Algorithms and the Making of Our Digital Selves*. New York University Press.

Cook, Amy. 2008. "Interplay: The Method and Potential of a Cognitive Scientific Approach to Theatre." *Theatre Journal* 59, no 4: 579–94.

Couldry, Nick, and Andreas Hepp. 2017. The *Mediated Construction of Reality*. Polity.

Crossley, Mark. 2019. *Intermedial Theatre: Principles and Practice*. Red Globe Press.

Damasio, Antonio. 2005. *Descartes' Error: Emotion, Reason, and the Human Brain*. Penguin.

Damasio, Antonio. 2010. *Self Comes to Mind: Constructing the Conscious Brain*. Pantheon.

Darroch, Michael. 2010. "Feedback Loop." 185. In *Mapping Intermediality in Performance*, edited by Sarah Bay-Cheng, Chiel Kattenbelt, Andy Lavender, and Robin Nelson. Amsterdam University Press.

Debord, Guy. 1995[1967]. *The Society of the Spectacle.* Translated by Donald Nicholson-Smith. Zone Books.

Dixon, Steve. 2007. *Digital Performance: A History of New Media in Theater, Dance, Performance Art, and Installation.* MIT Press.

Eskjaer, Mikkel Fugl. 2018. "Mediatization as Structural Couplings: Adapting to Media Logic(s)." 85–110. In *Media Logic(s) Revisited: Modeling the Interplay Between Media Institutions, Media Technology, and Societal Change*, edited by Caja Thimm, Mario Anastasiadis, and Jessica Einspanner-Pflock. Palgrave.

Farman, Jason. 2012. *Mobile Interface Theory: Embodied Space and Locative Media.* Routledge.

Farman, Jason. 2021. *Mobile Interface Theory: Embodied Space and Locative Media.* 2nd ed. Routledge.

Fischer-Lichte, Erika. 2008. *The Transformative Power of Performance: A New Aesthetics.* Routledge.

Freshwater, Helen. 2009. *Theatre and Audience.* Palgrave.

Freshwater, Helen. 2021. "Histories of Audiencing: On Evidence, Mythology, and Nostalgia." 37–52. In *The Routledge Companion to Audiences and the Performing Arts*, edited by Matthew Reason, Lynne Conner, Katya Johanson, and Ben Walmsey. Routledge.

Frieze, James. 2016. *Reframing Immersive Theatre: The Politics and Pragmatics of Participatory Performance.* Palgrave.

Giannachi, Gabriella. 2004. *Virtual Theatres: An Introduction.* Routledge.

Hansen, Mark B. N. 2004. *New Philosophy for New Media.* MIT Press.

Hansen, Mark B. N. 2006. *Bodies in Code: Interfaces with Digital Media.* Routledge.

Hansen, Mark B. N. 2015. *Feed-Forward.* University of Chicago Press.

Harvie, Jen. 2013. *Fairplay: Art Performance and Neoliberalism.* Palgrave.

Hayles, N. Katherine. 1997. "The Condition of Virtuality." 183–208. In *Language Machines: Technologies of Literary and Cultural Production*, edited by J Masten, P Stallybrass, and N Vickers. Routledge.

Hayles, N. Katherine. 1999. *How We Became Posthuman: Virtual Bodies in Cybernetics, Literature, and Informatics.* University of Chicago Press.

Hayles, N. Katherine. 2005. *My Mother Was a Computer: Digital Subjects and Literary Texts.* University of Chicago Press.

Hayles, N. Katherine. 2012. *How We Think: Digital Media and Contemporary Technogenesis.* University of Chicago Press.

Hayles, N. Katherine. 2014. "Cognition Everywhere: The Rise of the Cognigtive Nonconscious and the Costs of Consciousness." *New Literary History* 45, no 2: 199–220.

Hayles, N. Katherine. 2016. "Posthuman Cognition: The Power of the Cognitive Nonconscious." Keynote Speech at the Technology & The Human: Rethinking Posthumanism Graduate Conference. Waltham, Massachusetts.

Hayles, N. Katherine. 2017. *Unthought: The Power of the Cognitive NonConscious.* University of Chicago Press.

Hepp, Andreas. 2013. *Cultures of Mediatization.* Polity.

Hepp, Andreas. 2019. *Deep Mediatization.* Polity.

Hjarvard, Stig. 2013. *The Mediatization of Culture and Society.* Routledge.

Hjarvard, Stig. 2018. "The Logics of the Media and the Mediatized Conditions of Social Interaction." 63–84. In *Media Logic(s) Revisited: Modeling the Interplay Between Media Institutions, Media Technology, and Societal Change*, edited by Caja Thimm, Mario Anastasiadis, and Jessica Einspanner-Pflock. Palgrave.

Holub, Robert C. 1984. *Reception Theory: A Critical Introduction.* Methuen.

Jarvis, Liam. 2019. *Immersive Embodiment: Theatres of Mislocalized Sensation.* Palgrave.

Jenkins, Henry. 2006. *Convergence Culture: Where Old and New Media Collide.* New York University Press.

Jenkins, Henry, Sam Ford, and Joshua Green. 2013. *Spreadable Media: Creating Value and Meaning in a Networked Culture*. New York University Press.

Klich, Rosemary, and Edward Scheer. 2012. *Multimedia Performance*. Routledge.

Knowles, Ric. 2014. "Editorial Comment: Spectatorship." *Theatre Journal* 66, no 3: xi–xiv.

Krotz, Friedrich. 2008. "Media Connectivity: Concepts, Conditions, Consequences." 13–32. In *Connectivity, Networks and Flows: Conceptualizing Contemporary Communications*, edited by Andres Hepp, Friedrich Krotz, Shaun Moores, and Cartsten Winter. Hampton Press.

Lavender, Andy. 2016. *Performance in the Twenty-First Century: Theatres of Engagement*. Routledge.

Lévy, Pierre. 2001. *Cyberculture*. Translated by Robert Bononno. University of Minnesota Press.

Lewis, William W. 2017. "Performing Posthuman Spectatorship: Digital Proximity and Variable Agencies." *Performance Research* 22, no 3: 7–14.

Lewis, William W, and Sean Bartley. 2023. *Experiential Theatres: Praxis-Based Approaches to Training Twenty-First Century Theatre Artists*. Routledge.

Livingstone, Sonia. 2003. "The Changing Nature of Audiences: From the Mass Audience to the Interactive Media User." 337–59. In *A Companion to Media Studies*, edited by Angharad N. Valdivia. Blackwell Publishing.

Lundby, Knut. 2009. *Mediatization: Concept, Change, Consequences*. Peter Lang.

Machon, Josephine. 2013. *Immersive Theatres: Intimacy and Immediacy in Contemporary Performance*. Palgrave.

Massumi, Brian. 2002. *Parables for the Virtual: Movement, Affect, Sensation*. Duke University Press.

McCarthy, John, and Peter Wright. 2004. *Technology as Experience*. MIT Press.

McConachie, Bruce. 2015. *Evolution, Cognition, and Performance*. Cambridge University Press.

McCutcheon, Jade Rosina, and Barbara Sellers-Young. 2013. *Embodied Consciousness: Performance Technologies*. Palgrave.

Michelle, Carolyn. 2007. "Modes of Reception: A Consolidated Analytical Framework." *The Communication Review* 10, no 3: 181–222.

Murray, Janet H. 2017. *Hamlet on the Holodeck*. 2nd ed. Free Press.

Napoli, Philip M. 2012. "Audience Evolution and the Future of Audience Research." *International Journal of Media Management* 14, no 2: 79–97.

Nelson, Robin. 2010. "Experiencer." 44. In *Mapping Intermediality in Performance*, edited by Sarah Bay-Cheng, Chiel Kattenbelt, Andy Lavender, and Robin Nelson. Amsterdam University Press.

Pine II, B. Joseph and James H. Gilmore. 2011. *The Experience Economy*, updated edition. Harvard Business Review Press.

Rancière, Jacques. 2007. "The Emancipated Spectator." *Artforum International* 45, no 7: 270–81.

Rancière, Jacques. 2009. *The Emancipated Spectator*. Translated by Gregory Elliot. Verso.

Reason, Matthew. 2010. "Asking the Audience: Audience Research and the Experience of Theater." *About Performance* 10: 15–34.

Reason, Matthew, and Kirsty Sedgman. 2015. "Editors Introduction: Themed Section on Theatre Audiences." *Participations: Journal of Audience and Reception Studies* 12, no 1: 117–22.

Reason, Matthew, Lynne Conner, Katya Johanson, and Ben Walmsey. 2022. *The Routledge Companion to Audiences and the Performing Arts*. Routledge.

Salter, Chris. 2010. *Entangled: Technology and the Transformation of Performance*. MIT Press.

Sedgman, Kirsty. 2016. *Locating the Audience: How People Found Value in National Theatre Wales*. Intellect Books.
Sedgman, Kirsty. 2018. "Audience Experience in an Anti-Expert Age: A Survey of Theatre Audience Research." *Theatre Research International* 42, no 3: 307–22.
Sedgman, Kirsty. 2019. "On Rigour in Theatre Audience Research." *Contemporary Theatre Review* 29, no 4: 462–79.
Shirky, Clay. 2008. *Here Comes Everybody: The Power of Organizing Without Organizations*. Penguin Press.
Souza e Silva, Adriana de. 2009. "Hybrid Reality and Location-Based Gaming: Redefining Mobility and Game Spaces in Urban Environments." *Simulation and Gaming* 40, no 3: 404–24.
Stiegler, Bernard. 1998. *Technics and Time: The Fault of Epimetheus*. Translated by Richard Beardsworth and George Collins. Stanford University Press.
Walmsley, Ben. 2019. *Audience Engagement in the Performing Arts: A Critical Analysis*. Palgrave.
White, Gareth. 2013. *Audience Participation in Theatre: Aesthetics of the Invitation*. Palgrave.
Wiener, Norbert. 1948. *Cybernetics: Or Control and Communication in the Animal and the Machine*. MIT Press.
Wilson, Tony. 2008. *Understanding Media Users: From Theory to Practice*. Wiley-Blackwell.

1 Mediatization, Perception, and Experience

1.0 Foundations of the Mediatized Spectator

What is experience? How does one define it? How does one describe it? Can one do justice to the reality of another person's individual experience? Answering these questions are some of the challenges for researchers of spectatorship and audience behavior. At the macro-level, experience can be considered a perpetual act of being-in-the-world, or put more simply, people's evolving perception of the world they live within. At the micro-level of individual subjectivity and moment-to-moment perceiving, it is a phenomenological process that can only truly be understood in its fullest capacity during the act of experiencing an event. An experience forms through the perception of individualized micro-events collected into one cognitive and embodied response. The experience is then further defined by one's framing for meaning making and understood sense of reality, both of which are always fluid. One might attempt to recall the highs and lows of their experience, but this retrieval of experiential memory can never have the same impact of the event, or even fully capture the event. The meaning made through the experience also changes with time. As memories fade, new experiences create data points of comparison to correlate with, changing how one recalls and understands the original experience. Feedback surveys are often used to pinpoint specific elements of an audience member's experience, but again the data will be skewed in some manner or incomplete based on the individual's ability to understand all the intricacies of the original experience. Another complicating factor of researching experience is that individuals often are not aware of all the environmental and social influences impacting and determining how they experience the world moment-to-moment. Ultimately, experience is always in flux and being influenced by a series of competing factions that are social, cultural, and technological. While it would be impossible to explain all the networked influences determining audience behavior and expectations, this chapter aims to assist in establishing a framework for describing some of the influences of the technological.

This chapter lays out the foundation for analyzing various architectures of experiential spectatorship by explaining how mediatization and technogenesis

DOI: 10.4324/9781003398813-2

work hand in hand to propel a mode of intersubjectivity impacting people at the micro-level of individual daily experience as well as the macro-level of the broader social reality. Mediatization is engaged as a process that "*seeks to capture the nature of the interrelationship between historical changes in media communication and other transformational processes*" (Hepp 2013, 38 italics in original). Technogenesis refers to the process of co-evolution between humans and the technologies that define human "beingness" in any historical period. By showing how these interconnected processes shape people's ways-of-being in the world, a framework for evolving the ways of thinking about spectatorship and audience behavior emerges through a construct discussed as a *mediatized spectator.* This spectator is a subject whose ways-of-being in the world have been shaped by mediatization and technogenesis. Subsequently, their expectations about modes of exchange with(in)—with and in—performance/media events are impacted. The key here is that in order to best analyze spectator experience, one must first understand the many ways the spectator operates under processes of mediatization. Since today's social reality is filtered through the lens of mediatization, then all acts of spectatorship must be too. Hepp (2013) explains, "The traditional distinction made between experiences conveyed through media and those which are not becomes irrelevant, since the omnipresence of media outputs transforms both the individual and the social construction of reality" (21). The study of mediatization has arisen over the past 20 years with the impetus of answering questions pertaining to the role media and technologies play in shaping culture and societies. Unlike simple mediation and medium theory (McLuhan 1964)—which focus primarily on a specific media or technology at the micro-level—the term mediatization is offered to describe large-scale transformations within social spheres based on the interrelated forces of culture, society, and all media (Hjarvard 2013).

Within the fields of theatre and performance studies, the framework for mediatization is most often applied to performance as an adjectival framing denoting performance that incorporates media—filmic and televisual most often—or is mediated through some technological "medium." Phillip Auslander (2021) follows this train of thought to problematize the idea of liveness in contemporary performance. In the most recent edition of *Liveness: Performance in a Mediatized Culture,* he uses mediatization as a counter to the ephemerality of live in-person performance—one that requires a live performer and live audience in proximity—by comparing electronic and digital mediums such as film and recorded music alongside conventional concepts of live performance events such as music concerts, sports, and theatre. Auslander's update considers the role that technologies of mediated communication, like the internet, have in shaping experiential practices and spectators' roles within them. This is a brief departure from his previous argument. For example, he now explains how audio-performance walks, which form a new performance object of study compared alongside the internet, emerged as a form because "they engage with audiences whose expectations and cultural consumption bear the imprint of their experience with digital media, particularly with respect to what it means to be a spectator" (58). His inclusion of references to spectatorship in relation to digital media (Causey 2016; Harvie 2013; Lavender 2016; Oddey and White 2009) is useful, but he still focuses on the "ontology" of performance itself and a contested notion of liveness when relating to mediating forces. While Auslander now engages with Stig Hjarvard's (2013) theory of mediatization to support his evolving argument, his framing has more to do with the mediation of artistic forms and products versus a fully encompassing social framework for the formation of people's social realities. Instead of utilizing the concept "to indicate that a particular cultural object is a product of the mass media or media technology," (Auslander 2021, 4) I focus on mediatization on an intersubjective scale that covers the inter-relations between individuals, their mediated forms of communication, and the mediatized constructions of reality that come from these relations. As such, mediatization is utilized to analyze and understand the audience member's experience both prior to and during acts of performance, versus primarily focusing on an analysis of the performance itself.

This chapter helps to capture the essence of the mediatized spectator as a construct made through the forces of mediatization from a techno-sociological perspective. This perspective undoubtedly engages with a mode of technological determinism often decried by theatre and performance scholars deeply impacted by the liberal humanist traditions that brought about conceptions of liveness in the first place. This preciousness toward the face-to-face version of liveness is what Auslander's work pushes back against, specifically in an era where mediatized modes of performance have become ubiquitous and impactful. Mediatization acknowledges the impact of technologies on humans at both macro and micro levels while also showing how it is interconnected with other social forces that shape reality and is therefore not entirely a techno-determinist framework. By centering on mediatization, we are able to refocus on both the individual and communal experience of the audience member.

Mediatization should not be considered a completely universal framework for all social realities. It is useful, however, for developing a taxonomic framework for analyzing social actions where one can consider the influence of "meso-level" socio-technical forces or technics. Hjarvard (2013) explains that mediatization is also useful as it invites "cross disciplinary work in which we must seek to make use of theories and methodologies from a variety of disciplines" (4). This type of cross disciplinary ethos is crucial for the field of theatre and performance studies if it is to keep up with rapidly evolving technological paradigms. Through mediatization, a more holistic understanding of the influences of media and technologies on people and their social behaviors can be explored across fields of study interested in audience behavior. Mediatization theory allows one to break apart the divides within fields that might say one can only focus on specific objects of study.

For instance, theatre and performance can only focus on "live performance" or live performance that has been mediatized (i.e., intermedial, transmedial, multimedial) in some fashion. Instead, mediatization theory allows one to expand the horizon of work to focus on multi-modal forms utilizing the spectator in relation to the media frame as an object of study. Here, mediatization allows a focus on audience experience within multiple forms of media, across multiple fields, whether they are technologically constructed or not. Remember, in this construction, *all forms of spectatorial events are media.* With audience experience becoming the central focus, one must first consider how that audience member came into being as a mediatized spectator, which means focusing on their intersubjective relationship with their technological objects and media forces.

The following section lays out the basis for mediatization and how it acts as a social force that impacts the way a human being situates itself within a perceived reality. Next, the concept of technogenesis is covered to explain how specific *technics* impact the way these humans perceive the world around them, allowing the formation of a reality. Nested within specific waves of mediatization are unique technics that operate as technologically derived social forces informed by specific types of media and technologies that work alongside other social forces (i.e., politics, religion, culture, sports, performance) to shape and reshape people's everyday perceptions of reality and their actions derived from those perceptions. Technics operate at a meso level, situated inside larger waves of mediatization such as mechanization and digitalization. These technics act as middles between people and their social environments and are in constant flux and tension with other technics. In the age of digitalization, these technics mediatize all social systems and act to expand the concept of mediation to both frame communication as well as act as conduits of communication via digital technologies and digital media. Each technic imparts a specific form of relationality as operational logic that is harnessed within the four *Architectures of Exchange* as frames for spectatorship.

After describing the ways both mediatization and technics impact one's sense of being in the world, I turn to a phenomenological perspective to better explain how a contemporary spectator becomes mediatized through the relational interactions of their *perceptual apparatus,* another construct that combines the perceiving capacities of one's mind, body, and environmental senses. This top-down approach serves as a meta-lens for analyzing the various experiences a mediatized spectator might have within the various architectures. To best understand the overarching structures, the operations of these architectures as frames for exchange, and the audience's experience with(in) each architecture, one must first expand outward toward the macro-process of mediatization and then drill down layer by layer, frame by frame, to show how technologies and technics are changing people at the level of individual subjectivity which then manifests itself through evolving audience behaviors, expectations, and experiences.

Intersubjective Frames of Deep Mediatization

The social world is one that fundamentally constructs the reality people perceive and experience, and this social world is itself constructed through a combination of mediated processes described as mediatization. Neither the individual nor the social world are pre-given constructs brought into existence prior to an interaction between the two. This has always been true. Contemporary social worlds and the individuals with(in) them have co-evolved to rely on specific technological forces and tools for their construction. As such, one must understand the interplay that occurs through processes of mediatization and society, and specifically the impacts of deep mediatization. Likewise, to understand the many ways in which the contemporary spectator performs its role with(in) performance events, it is crucial to first understand how they, as individuals, operate as social actors with and in today's mediatized social world(s). These social actors are constantly performing as part of distinct but overlapping media cultures. Andres Hepp (2013) defines these media cultures "as *cultures whose primary meaning resources are mediated through technical communications media*, and which are *'moulded' by these processes in specifically different ways*" (70, italics in original). As I will argue, individual technics operate as the defining meso-processes paired to distinct media cultures occurring over the last 40 or so years. These cultures do not have clear breaks between them as each often builds off the previous in certain ways and also remains in parallel to other media cultures. Each, however, is defined as belonging to a larger historical macro-process of mediatization.

The overarching structures of mediatization are enacted over large historical periods defined as "waves of mediatization" by Couldry and Hepp (2017). These temporal waves contain specific technics based on individual technological innovations and relations. The most recent waves have brought about what is described as deep mediatization, a force where connection to the effects and affects of media becomes ubiquitous (Hepp 2019). Mediatization encompasses **all** technological forms of mediated communication—commonly referred to as media—that impact daily perception **and** the technologies that drive this communication.

> The social world is the intersubjective sphere of the social relations that we as human beings experience. Those relations are rooted in everyday reality, a reality nowadays always interwoven with media to some degree. The social world is, in turn, differentiated into many domains of meaning, even though it is bound together by multiple relations of interdependence and constraint.
>
> (Couldry and Hepp, 20)

For example, digital media and digital technologies are interconnected tools dictating the very nature of perceived social realities through a wave described as *digitalization*, which is itself a large-scale technic operating as an individual historical/temporal wave within mediatization. Digitalization is mentioned here first as it is the prevailing wave encompassing the modes of experience described in this project and covers many of the technological innovations of the past 50 years. Within digitalization are multiple technics attributed to sets of technologies and media that drive particular modes of experiential exchange. The interrelations between these individual elements inform how people perceive and operate in today's digitally augmented social world.

Mediatization is therefore a process whereby social and cultural relationships become mediated by a techno-social paradigm that drives social change dictated by the underlying "infrastructures of communication" (38). Mediatization operates in constant flux as a system of technological and human interactions. The multiple forms of media performing in the operation of mediatization are referred to "both as technologies, including infrastructures, and as processes of sense-making" that form the social reality of the current world (5). People's understanding of reality is based on "material process (objects, linkages, infrastructures, platforms) through which communication, and the construction of meaning, take place" (3). Therefore, the nature of the social is one in which the interrelated processes and (inter)actions between all media mold perceived reality. This reality is configurable and in flux based on the inputs added to the fabric of the social via mediated communication and the outputs of people acting within this fabric. This social is itself constantly changing and is not a simple pre-constructed frame in which people act; it is always shifting based on the relational capacities of people, technologies, and cultures. Mediatization helps explain the relationships between changes in media and communication and the changes between culture and society which then filter down to everyday behaviors and actions of individuals. It works as an inter-related process built out of a dialectic relationship with human subjectivity versus a simple unidirectional cause-and-effect structure or simple techno-determinist application one might focus on in medium theory. Considering mediatization as a dialectic shows how media and communication are inseparable from culture or society because both are integral to the internal *and* external structuring of social worlds and the realities defined by those worlds (35). The social world is therefore grounded "not in ideas, but in *everyday action, that is, in practice*" (21, italics in original). The formation of social worlds is therefore an ongoing performance of (inter)actions between all elements contained within and is defined through the experiences of individuals within these social worlds.

Once a social world has reached a point where mediatization can no longer be removed from the social, and the social can no longer be understood divorced from mediatization, a paradigm of deep mediatization occurs. Deep mediatization is both a condition and a "meta process involving, at every level of social formation, media-related dynamics coming together, conflicting with each other, and finding different expressions in the various domains of our social world" (215). Deep mediatization has emerged as an advanced stage within the wave of digitalization where virtually all aspects of daily life, including the infrastructures of the social sphere, are influenced and connected to digital media (Hepp 2019, 5). By understanding deep mediatization as a meta-process constantly in flux, one can see how the varied set of technics within a specific wave inform and define overlapping sets of realities and the subsequent ways of operating in those realities. The analytical frame offered in this project shows how these individual technics emerge and inform specific overlapping modes of audience experience within digitalization and datafication.

The recent history of humanity can be divided into four distinct waves of mediatization. Each wave represents a "fundamental qualitative change in media environments sufficiently decisive to constitute a *distinct phase* in the ongoing process of mediatization" (Couldry and Hepp, 39, italics in original). The waves are experienced in various levels or strength based on contextual information such as cultural, political, geographical, and economic determinates. Each wave marks an origin for social reconfigurations based on changes in media ecologies, but unlike medium theory, which posits that individual media/technology dictate change as a quick paradigm shift, waves of mediatization take on a more broad-scale approach and evolve over time with visible overlaps. The four waves identified by Couldry and Hepp include *mechanization, electrification, digitalization,* and *datafication.* As one will see, the operations of technologies in each wave are not necessarily doing something novel; they often remediate existing modes of mediated communication. To use a crude example, oral language (a technology) remediated non-verbal language and pictorial forms of storytelling. Written language then remediated oral forms, mechanized print remediated handwritten language, and so on. The focus within the four waves of mediatization becomes the way discrete technological innovations merge based on their base construction and mode of transmission. Utilizing these individual waves as large-scale technics, one could embark on a study of audience behavior based on each social-temporal frame. The focus of this project is primarily on digitalization due to its pervasive impact driving deep mediatization and the relational qualities of interactivity found in experiential modes of performance. Before jumping in headfirst, it is useful to briefly explain the four primary waves as a primer for how audience studies have often focused on narrow aspects of medium theory and the unidirectional nature of reception theory.

Mechanization began with the printing press and included all the technological adaptations that allow simple machines to replace human operated modes of communication, such as handwriting and oral storytelling. While the printing press transformed the way written language was produced, disseminated, and consumed, it was not the only process of mechanizing language. Tools such as print blocks had existed for centuries. It, however, initiated a large-scale evolution in societies because it allowed written material to reach many more human beings in technologically evolving societies than was previously possible. It also allowed multiple formats of mass-produced text, including the broadsheet, pamphlets, and eventually newspapers. The sheer amount of new literary material and formats sparked a surge in innovation while changing the ways people interacted socially. Mechanization also spread across other fields beyond communicative media. This spread includes across mobility and production through the development of steam and then combustion engines that propel vehicular travel (train, automobile, plane) and the automation of factory work. These fields changed cultural systems through modifications to geographical constraints and economics as part of industrialization, while printed text changed social systems through changing dynamics in education and communication. As the printed word spread around the world, higher levels of literacy came with it, leading to more societies entering an advanced state of increasing social complexity. The explosion in both the form and content of printed media allowed for a stratification of audiences and communities. Genre and form paralleled migratory patterns propelled by technologies of movement. While the train, as a mechanized form of travel, came late in the history of mechanization, it propelled the mediatization process into a higher gear. Mass migrations led to urban booms and transcultural intermingling which then influenced the formation of national states trying to solidify their identities with mediated communication assisting. The prolonged process of mechanization continued into the late twentieth century, with its latter stages introducing the mechanization of visual and auditory art forms that include photography, the phonograph and gramophone, and the stereoscope. Each of these inventions would quickly be remediated through *electrification.*

Beginning in the early nineteenth century and reaching its apex in the late twentieth century, *electrification* accelerated the pace of changes in social worlds. Electrification describes "the transformation of communications media into technologies and infrastructures based on electronic transmission" (44). Electrification augmented many mechanical processes, but when discussed relative to communication, it is primarily identified with the telegraph, telephone, film, radio, television, and the complex networks created between each of these media. The telegraph, telephone, and radio were first thought of as horizontal communication systems that allowed reciprocity in modes of communication between two involved parties. The reciprocal capacity quickly became "sender-centered" and primarily unidirectional when they matured and became the domain of political and corporate entities in the late 1920s. The unidirectionality of electrified media became commonplace through the mass adoption of film and then television. This intensified through the emergence of targeted marketing research. The telegraph and then the telephone allowed communication between vast geographical distances to occur that may have taken months in previous eras. Through expansion and deepening, electrification allowed modes of distant communication to become near instantaneous, changing how people understood and constructed concepts of time, place, and space. Electrification offered a paradigm shift through the interlinking of the various media and the infrastructures that allowed their transmission. The electric grid allowed broadcast networks to form and morph into extensive cable networks and then entire media conglomerates. These conglomerates were intricately linked to geographic, political, economic, and social institutions, which expanded as electrification deepened and spread. Though electrification is often tied to the first stages of globalization and the flattening of distance and temporality, it was not a process that spread equally across all parts of the world and all social spheres. Much of the audience studies of the middle to late twentieth century is based on the technologies of electrification and thus focus more on concepts of reception and medium specificity and less on relationality and feedback mechanics found in digital media that are predicated on user interactivity.

The next wave of mediatization *digitalization* serves as the most visible socio-technical process impacting people today. Digitalization began in theory in the 1940s, but it was not until the last decade of the century that it firmly took root. In digitalization, the primary mediator of human communication is the digital sphere and its many individualized domains contained in that sphere. The backbone of the digital sphere is the complex network of multimodal communication labeled the internet. Often called the World Wide Web, the internet began as small digitally connected networks, mainly between academic and governmental entities. The substantial leap to world connectivity came in the late 1980s with the creation of programming languages that would allow "metadata" and "hypertext" to form a largely unified web of electronic files. Through new protocols of computer-to-computer communication, a networked digital infrastructure with the capacity to connect all parts of the world and all people within that world emerged. The first user-friendly websites were developed in the early 1990s, and this led to the need for web applications that could effectively navigate the linkages. The creation of multiple internet browsers and search applications by corporate entities such as Netscape, AOL, and Google changed how communication operated and recirculated via digital and virtual systems. These changes led to innovative new applications that transformed how previous media were used. Each step in the development of the internet changed how human communication worked and changed the social systems that dictate how humans operate as sentient beings. Couldry and Hepp explain,

> The result of these cumulative and interlocking steps is a strikingly complete transformation of 'the internet' from a closed, publicly funded and publicly oriented network for specialist communication into a deeply commercialized, increasingly banal space for the conduct of social life itself.
>
> (50, quotations in original)

Under digitalization, the internet quickly went from being a tool for institutional peer-to-peer communication and transformed into a mediatized platform for exploring, creating, and experiencing all of reality.

Digitalization also allowed the formation of multiple domains of technologically framed social media which act as multifaceted technologies that "comprise platforms which, for humans, literally *are* the spaces where, through communication, they *enact* the social" (2, italics in original). Under digitalization, social worlds are no longer created primarily through face-to-face communication; instead, digital media have become equal partners in the construction of people's social reality. This construction is becoming increasingly reliant on digital processes and digital technologies. Similar to my anecdote at the beginning of this book, the impact of tablet technologies on childhood understandings of the world is used as an example by Couldry and Hepp. When a child interfaces with the world through an internet-connected tablet at an early age, the technological tool becomes part of their socialization process. For this child, a sense of perpetual "mediated connection is now installed from early on as an operation condition of 'the world' itself" (150). The child becomes socialized and habituated to understanding the world through the lens of their digital devices. As they grow up, they feed that understanding back into every aspect of socialization they have with others, creating an increasingly powerful feedback cycle of mediatized communication and interpersonal relationality. Their very sense of being, *of a self*, is reliant on the forces of mediatization brought about by their early and continual connection to digital tools and logics.

The interdependence of our completely mediatized environments causes a reconceptualization of what it is to perform one's own construction of selfhood. Before digitalization and the most recent wave, *datafication*, the performance of selfhood was predicated on analog interactions with the rest of the world primarily through communicative acts of language (both verbal, textual, and embodied) as processes of social performativity. Digitalization and datafication have transformed the site of the self from a performance where, "being 'someone' shifts from being associated with a certain quality that self and others can abstract from the stream of habitual action to being a continuously managed 'project,' that is, an 'external' responsibility of the self toward the social world" (146, quotations in original). Put more simply, individual selves have become mediated by the digital platforms now defining their new social realities via the processes of mediatization. These multiple selves exist in relation to the intersubjective connections between the digital sphere and the public sphere of everyday offline life. These selves are now dialogical selves, beings that exist in constant networked communication between their corporeal and digital identities. The performance of selfhood is therefore constantly mediated through the digital, becoming increasingly more "processual," causing a necessity for understanding "the material processes of forming and sustaining a self" within deeply mediatized social spheres (148). With selfhood transformed through mediatization, individual acts of spectatorship are also transformed, offering the concept of the mediatized spectator.

Waves of mediatization fundamentally change the way social worlds are formed and how individuals perceive both those worlds and their place in them. Twenty-first century technologies of digitalization are increasingly becoming the all-encompassing bounding forces forming all social structures that people interact with. These interactions construct new systems of perception and reality that require a "material phenomenology" when approaching technologically mediated social realities (5). In this approach, the world is made up not only of people's subjective interpretations of events and experiences but also the effect of the "things" that are part of those subjective interpretations as well as the "things" themselves. Media and mediatized technologies have crucial agency to impact the way people perceive and therefore have agency to change their interpretation of the very things they perceive. A major part of the argument presented in this book is that mediatization effects and affects one's perceptual apparatus through technogenesis and causes the performance of spectatorship to evolve from something less akin to simply receiving but more like experiencing through multiple modes of cognitive, affective, and environmental relationality.

When one's sense of social being-in-the-world becomes aware of their own implicitness in the interactive making of that world, they become habituated to have a greater stake in the making of a performance from within as its mediatized spectator. This ushers in a personal need to materially engage in ways that are measurable or tangible or, put in a more encompassing terminology, experienced. This seems tautological but is necessary to understand as the very feedback loop that constructs meaning for people and thus their social realities. These realities are continually reshaped by contemporary technogenesis, and this reshaping impacts both people and the social worlds containing those people. Technogenesis exists as a continual process of change that is itself impacted by various technics inside waves of mediatization. Contemporary forms of technogenesis represent the interconnected changes brought about by digitalization and datafication and become the backdrop for an expanded notion of spectatorship that encompasses multiple domains of research from various fields, all of which are interested in developing methods for understanding contemporary audience experience.

What Is Technogenesis?

Can you think of a time before the internet? If the reader belongs to Gen X like I do, you should be able to. If so, can you remember what it was like to imagine beyond your immediate social sphere? Do you remember how slowed down time seemed? Do you remember what it was like to have to memorize phone numbers and not be subjected to constant notifications? Now think of your life today. How has constant connection to digital technologies changed your perception of the world and how you go about living in it? You have changed as technologies have evolved and become more embedded in your everyday life. This change is called technogenesis and has been a constant aspect of all evolutionary social change throughout human history.

The term is often credited to philosopher Bernard Stiegler. In his three book series *Technics and Time* (1998, 2009, 2011), he engages with Heidegger's phenomenological term *dasein* (the to-be/to exist), the transcendental humanism of Rousseau, and the origin of humankind in relationship to technologies as explained by Gilbert Simondon. In book one, *Technics and Time: The Fault of Epimetheus,* Stiegler (1998) uses the term technogenesis to tease out a construction of humanity as one that is forever and inextricably linked to both technological progress and the technics that encapsulate and dictate culture. Adopting Simondon's definition of technics as a "process of concretization" (22) that is forever in flux, Stiegler explains that once the human species enters an age where it is impossible to delineate causality of effects between an individual and an individual technology—because of the simultaneous and infinite multiplicity of interactions between many technologies and many individuals—technics become the prevailing logics defining what constructs the human species as a unique socio-technical animal. Philosophically, a technic adopts the logic of both *episteme* (deferential knowledge for knowledge's sake) and *tekhne* (functional knowledge created through the application of epistemes) to create a system of knowledge and beingness that is forever in process, forming the horizon of all possibilities for human existence (74). If technologies are the tools that humans interact with, technics are the overarching logics that those tools inspire. They are tool-like in the way they help shape the potential of humans and how they can also be wielded by people in physical and symbolic ways. Both metaphysics and materiality combine to create technics as assemblages that allow a person to operate as a technical individual; one whose beingness conjoins with a technical object's symbolic operation and/or purpose. Technics are what make human beings distinct from other animals as they are uniquely connected to their social life. In this paradigm, a technological apparatus surpasses simple utensil, or tool, to become an agent—object/subject with agency—in systems of reality accessible by people via their perceptual apparatus.

Technics operate similarly to the waves of mediatization described in the previous sub-section. In fact, waves of mediatization can be thought of as historical macro-technics. They are large-scale technological processes that work with and on people and the social systems created out of the relationship between the two. These systems are also always in flux, and as technologies change over time—often more quickly than human beings do—they exert a force that causes them to enact a social and cognitive evolutionary process with and on people. That process is what Stiegler and subsequently N. Katherine Hayles (2012) discuss as technogenesis: The ongoing relational process between a human being's perception of self and the technics and technologies that bring that self into being as a mediated construct. Utilizing technogenesis to frame one's understanding of their being-in-the-world requires a shift in thought away from linear causality and beliefs in an origin or stable construct of innate human be-ingness (ontology) toward an understanding of ongoing relationality and interconnectedness. This shift allows a slight remove from a completely techno-deterministic perspective as there is no originary object that the technology determines; instead, the object (a person) and the technology move together, impacting each other along the way. Stiegler and Hayles use the terminology co-evolution to explain the process. As such, one engaging with it must suspend propensities toward cause and effect, instead considering operation and process as functions inside dynamic and unending systems of relational exchange.

Technogenesis is therefore a fluid process of change and adaptation between a person—or sets of people in a social frame—and all the technologies they interact with and relate to alongside specific technics. In Part I of Stiegler's book, titled "The Invention of the Human" (1998: 21–134), he marks the beginning of technogenesis as an interpretable phenomenon by citing Karl Marx's materialist orthodoxy on history. He uses Marx's theory of interpretation to explain technology's relationship to both human nature and human culture: "Technology reveals the active relation of man to nature, the direct process of the production of his life, and thereby it also lays bare the process of the social relations of his life, and of the mental conceptions that flow from these relations" (26). While Marx discussed the material impact of human social systems as the primary element that builds history, Stiegler goes further to argue that both the foundations of humanity and humanity's perception of time emerge from the material relationships among technology, history, culture, and an individual's own perception of selfhood. That self is constantly forming and reforming based on each of these social forces. Stiegler's utilizes technogenesis to explain how time and temporality are technical constructs dependent on the inter-relationship between human beings and technics.

Hayles (2012) engages with technogenesis to explain how technologies fundamentally alter the ways people think, perceive, and operate in the world. Hayles (2014) further explains, "humans have always been inte-grated into their environment and have co-evolved with it" (98). What comes first though, the subject or the technology? Neither. They are inter-dependent on each other's influence. As part of the social and technical environment, technics allow the construction of the mediatized subject possible, while that subject allows technologies to operate and evolve in a manner that creates the symbiotic link. This is the basis of Simondson's concretization (Iliadis 2015). This tether is always in tension, however, due to concretization, where the mechanic evolves to take on more agency which in turn forces the human on the other side of the pairing to adapt, regaining some of that agency. A person constantly negotiates its own power to make sense of its place in the world and its relationship to other agents within a network of relationality.

Technogenesis is also an ontogenetic evolutionary process whereby all technological tools formulate the nature of an individual's interactions with the world (Hayles 2012). Ontogeny is used by Hayles to explain the combined structural changes to the cognitive, physical, emotional, and social aspects of a person based on environmental factors as opposed to hereditary genetic factors. Ontogenesis is a developmental process of combined epigenetic changes that develop rapidly and imprint upon a person's perceptual apparatus in ways that can be passed on to subsequent generations. For example, prolonged use of the smartphone makes certain morphological and cognitive changes in the human brain, which then causes both social and cultural adaptations to ensue (Potzch and Hayles 2014, 98). These adaptations pass on to the next generation. The relationship between a person and a smartphone initiates a change in relationship between that person and the social, cultural, and technological worlds they inhabit. People's use of tools (technology), and subsequently a technology's use of people (technic) changes how the world is perceived and interacted with those people. In this manner, evolutionary change should never be solely considered a process of forward momentum, marching forward toward a "perfect" model. Instead, it is an ongoing process that can take steps sideways or fold back on itself to propel adaptation. When considering technics and technogenesis as processes involving technologies—through the use of, perception through, and interaction with—it shows how one can frame experiential spectatorship as a set of evolving modes of exchange brought about through adaptations in the perceptual apparatus of the mediatized spectator.

"Hyper" Text as Technology and Technic

In the following pages, technogenesis is grounded in mediatized culture and networked behavior using examples of digitally informed reading as a way of breaking the unidirectional model of reception in favor of perception and relationality, which are integral to experiential spectatorship. In 1999, Hayles wrote,

> Different technologies of text production suggest different models of signification; changes in signification are linked with shifts in consumption; shifting patterns of consumption initiate new experiences of embodiment; and embodied experience interacts with codes of representation to generate new kinds of textual worlds. In fact, each category—production, signification, consumption, bodily experience, and representation—is in constant feedback and feedforward loops with the others. (28).

Hayles' statement about text as technology is helpful when considering technogenesis. Language conditions people's actions and modes of perception.

Language has operated as a meta-technology in many instances. Text and the alphabets that form it condition human thought. Many of the philosophers and social theories of the postmodern turn focused their pursuits on this strand of technogenesis without explicitly naming it. Since the development of written text, human beings have structured their idea of reality based on the formal constraints and rules of the medium through which text surfaces. These constraints are historically and culturally specific, however. For example, in the English language, printed text is read from left to right on a page using individual letters that form words. In other languages, text sometimes moves in different directions, and words are comprised of complex symbol combinations shaping different modes of relationality. Formal rules of written language did not exist during times when communication worked primarily through oral traditions. As written languages were adopted by the educated elite, a more concrete form of rationality began to establish itself as a mode of social life with its own codified rules and frames for reality. These rulesets would help establish hierarchies of power and hence frame even further understandings of one's place in the world. With the invention of the printing press, these rational rulesets began to become the standard from which all meaning-making occurred. Reading based on linear rules became ingrained in Anglo/European social systems and remained the dominant mode until the introduction of the internet. In roughly the past 30 years, the social condition established after the mass adoption of printed material has slowly eroded due to the introduction of an entirely new way of developing knowledge: Relational interaction as found in the logic of hypertext mediated language (Hayles 2012).

The internet shapes the way the human brain processes information and how that information then conditions the way people behave. Nicholas Carr (2010) takes up this argument by discussing how the human brain rewires and re-networks itself based on the experiential information encountered and processed. This physical rewiring then allows a symbolic rearranging of social spheres and social relations. While the influence of the internet is Carr's primary focus, he takes time to describe other technological influences such as the development of maps, newspapers, clocks, and computers; each offering new ways of extending or supporting people's system of sensing and comprehending the world around them through the way they read information. These communicative media all operate as *intellectual technologies* that enact a "set of assumptions about how the human mind works or should work" (45). Intellectual technologies propel technics that then form the material conditions for how people's minds shape their everyday actions. Carr explains, "it is our intellectual technologies that have the greatest and most lasting power over what and how we think. They are our most intimate tools, the ones we use for self-expression, for shaping personal and public identity, and for cultivating relations with others" (45). This focus on intellectual technologies brings forth questions of communication and media specificity related to meaning-making.

Writing, as a text-based construction, is one of the historically most influential intellectual technologies to impact human minds and human civilization. Starting during the Enlightenment, sustained concentration on long-form textual information gradually became the dominant mode of information dissemination, impacting the way European societies perceive and interact with the world. Text in the form of a written book forces the mind to slow down and develop a more singular and narrow form of focus. Prior to the mass adoption of printed text, most minds were programmed to react like they did in the wild, where scanning the surroundings and being aware of everything around oneself helped maintain safety. Prolonged exposure to this new form of reading information overtime began to imprint a mode of perceiving the world with a specific structure and manner of reception. The written format of books operated as the ideal medium and technology of information transfer for over 500 years, imposing a technic of literary thinking based on a form of linear causality mimicking the format of the technology itself. As such, it informed one of the most powerful technics through which human beings could adapt and evolve. It is helpful to consider the printed book as a technology that informs experiential modes of human subjectivity formation and human selfhood. Reading a book on a printed page or on a static digital reader for that matter allows one to deeply immerse oneself within the lines of the page as they also immerse themselves within the content. The form in which content appears within/on coaxes the reader into a narrative or informational realm within the imaginative boundaries of the page-space, keeping the reader satiated without the urge to stray outside of its linear confines. The linearity of the book format helps promote causal efficacy and rationality as dominant modes of human beingness entrenched in the liberal humanist model. The liberal humanist literary subject is itself a technogenetic construction developed through the interconnection between linear print technology and human cognition. This technogenetic condition is gradually being replaced by a non-linear paradigm via the internet.

With its hyper-textual, multisensory, and experiential format, reading via the internet has supplanted the logics of printed text, becoming a culture changing intellectual technology due to its structure as a multidirectional form of communication. That multidirectionality supports an ever-present need for relationality in all experiences. The internet, as textual/visual/communicative technology, is ever-present today and has fundamentally changed people's perception of the world and, thus, the way they learn. As a dominant space and medium for information exchange, it helped create a new technic ushering forth multiple epigenetic changes in humans. These changes are part of the defensive mechanism ingrained in the biological process of neuroplasticity. Neuroplasticity allows localized and encultured forms of cognitive evolution at the neuronal level and operates as a basis for human learning. Not hardwired genetically, the physical construction of our brains is "constantly changing in response to our experiences and our behavior, reworking their circuitry" (31). This perpetually changing nature of the brain in relation to technologies and technics is the focus of Hayles' (2012) use of technogenesis. She explains:

> As digital media [...] become more pervasive, they push us in the direction of faster communication, more intense and varied information streams, more integration of humans and intelligent machines, and more interactions of language with code. These environmental changes have significant neurological consequences, many of which are now becoming evident in young people. (11)

Young people are not the only ones affected by these technologies, but they are a useful example of large-scale technogenetic evolutions brought about by digitalization (Lewis 2018). Relying on the Kaiser Foundation's "Generation M"[1] research study on media use (Roberts, Goehr and Rideout 2005), Hayles (2012) states that the cognitive shift accelerates the younger the cohort (69). Computer-assisted adaptations to textual information, in the form of multimedia and hyperlinked text delivered over the internet, have culturally molded today's typical American college-age student (Rosen 2011; Twenge 2017). The internet has also become the operational backdrop of all convergent technologies such as smartphones and mobile media (Farman 2012, 2021). It has also allowed the rise of datafication which promotes algorithmic functions to create a powerful new paradigm of control and relationality with people. The way the framework of the internet has become embedded within nearly all contemporary technologies accelerates the cognitive shift due to their pervasive interconnection with a human's sense of being-in-the-world. Today's technological devices instigate monumentally impactful technics, which "enables new feedback loops and new forms of amplification between human evolution and technical developments" (Potzch and Hayles 2014, 98). The rapid adoption of new technological materials amplifies contemporary technics.

When raised with and culturally habituated to hyperlinked text and multimedia delivered via web interfaces, a reader's ability to stay immersed in linear text recedes. It is replaced by interaction as a multi-modal form of relationality. One based on simultaneity due to the engagement of multiple senses at once. Reading through multiple senses then ushers forth a necessity for deeper experiencing. Just like reading, spectatorship—as a process of perception and interpretation—is adapting to hyper-textual and hyper-mediated social worlds. These worlds require us to think in terms of a matrix or assemblage with interlocking agential points of information as opposed to a point A to point B mode of thinking required in an uninterrupted line of text. Hyper-textual reading keeps readers engaged by urging them to constantly jump from place to place and portal to portal in a process of spatial-visual-cognitive multitasking. This form of reading and knowledge creation is based on hunting and gathering information bits and correlating them against other bits. Hyper-reading is the norm in digital contexts such as blogs, wikis, social media feeds, news sites, etc. where readers employ tactics of keyword filtering, "skimming, hyperlinking, 'pecking' (pulling out a few items from a longer text), and fragmenting" (61, quotations in original). Hayles adds to this list *juxtaposing* where readers have multiple texts open on the screen at one time. Just think of the frequent practice of having multiple browser tabs open for ease of information access. As hyper-reading becomes a primary mode, it neurologically imprints on readers a hyper-active and relational perspective (Rosen 2010, 2011). These readers become figuratively addicted to multitasking and constant user-initiated interaction. This "addiction" has rapidly become a norm and may soon completely replace the historically young tradition of deep learning promoted by the unidirectional information received via text in a printed book. Because of the current mediatized environment, people's minds are adapting into deeply hyper-active modes of perception. Eventually, slow reading and sustained concentration may no longer be necessary to navigate one's social worlds. The speed of contemporary technogenesis is fundamentally changing the way today's human beings read the world. By engaging human beings in increasingly interactive processes of exchange, hyper-attention impacts the ways they receive all modes of media. If today's readers have adapted to hyper-textual and a highly interactive world, then it is logical to think that their whole perceptual apparatus is changing to develop new ways of spectating.

The Perceptual Apparatus of the Mediatized Spectator and Its Posthuman Analogue

Imagine your body. What does it look like? What does it do? How does it do what it does? In what ways does that body interpret the world around it? What tools does it access and connect with to conduct that process of interpretation? How is your body an extension of your mediatized environment? Your body is part of a perception machine; *a perceptual apparatus* (Barad 2003; Hansen 2006, 2004; Massumi 2002; Chandler and Munday 2011a, 2011b) made up of material elements (skin, tissue, organs, nerves) and symbolic forces (ideologies, textualities, semiotics) always in relation to environmental factors (cultures, technologies, spatialities) through sensing, affect, and perception. This apparatus works to process both material and immaterial inputs and outputs via sensory reception, affective response, and relational processing called cognition.

The perceptual apparatus puts into constant relation one's body, their mind, and the technological environment they are embedded with(in) to sense and interpret the experience of an individual's being-in-the-world. This follows Brian Massumi's (2002) conceptualization of change that is not initiated from "in-between" as a "middling being but rather the being *of* the middle—the being of a relation" (70). In this instance, change is the constant state of existence for an individual which enacts its beingness through perception, a step in the process of experiencing. Experience is a process with perception and thought working on two sides of the same continuum, both working through relationality (91). Without naming mediatization, Massumi offers the basis of this relationality on the technological forces/structures of postmodernity, specifically those of the digital and the network as epistemological constructs:

> Its network is what connects coding to coding, codification to codification, and each to its own repetitions in an ebb and flow of potentialization-and-containment. The network distributes. Interlinks. Relates. The network is the relationality of that which technologies *give body to relationality as such* and set into motion. (87)

The perceptual apparatus is therefore the machine and conduit which operates in a constant state of potentialization, waiting to apprehend that from which it will be constructed while constructing the reality it will apprehend with(in). Within deep mediatization, that reality is always forming through digitalization, offering a mediatized spectator who is never stable in its construction but also always offering a new set of potentials based on the frame through which it is viewed externally.

For the mediatized spectator, the operation of experiencing the world is a form of embodied cognition where consciousness forms from the interconnection between the brain, body, and environment (Damasio 2005, 2010; McCutcheon and Sellers-Young 2013; Varela 1991) forming the perceptual apparatus. Embodied cognition comes not from any one element but from all these interacting elements and allows a continual formation of one's world and their place in that world. One's world is itself "enacted" through the feedback processing of the brain and body sensing capacities through perception. Perception, however, is not an external process divorced and "waiting" to sense the world. The world is created through the act of perception as a process of enactment (Varela 1991, 174). Embodied perception exists through interconnected relational processes of the perceptual apparatus and is therefore always a networking of embodiment, cognition, and environmental sensing.

Unlike the Cartesian formula of *I think therefore I am,* consciousness and reality exist in simultaneous and equal actions of *sense, think, relate.* The *I* (as subject) forms out of these interlinking processes that make up perception. Perception, as such, is an operation in which the body simultaneously senses and apprehends material conditions via affect at the same time the brain processes what the body apprehends to develop conscious and unconscious awareness of the world (Chandler and Munday 2011a, 2011b). The body and brain are not independent; they operate with the environment in a continual and nonstop relational feedback loop. To grasp the complexity of the perceptual apparatus and its relational processing, one must accept an ecological approach to perceiving. As explained in the previous sections, the environment through which this processing occurs is a mediatized one and infused with a plethora of technics and technologies. Perception and interpretation are part of a continuum where communication technologies act in concert with the feeling body to perform as a mediating influence in the relational process. This process is dual-directional in that it both receives and transmits information simultaneously. The process of transmitting and receiving multiplies as a body connects to its environment as an informational source and receiver, and the brain as signal processor.

The body, brain, and environment all act as conduits through which information passes in *two directions at once*: To and from the other elements in non-stop relationality. Without the body, however, the other two could not connect or relate, which allows meaning-making. This means that our social and performative world is made possible "through the capacities of our bodies"; it is through various "sense-making practices" that we can construct and conceive of our technologically mediated world (Couldry and Hepp 2017, 18–19). The perceptual apparatus is therefore a relational object, a perceiving machine operating to create meaning through the many relational activities it experiences. Thinking of mediatized spectators as subjects whose mode of experiencing performance is based on the relational aspects of their perceptual apparatus allows one to conceptualize modes of spectating as acts of embodied mediation. The perceptual apparatus works as a machine for sending and receiving information that helps to create meaning via exchange embedded in frameworks of experiential spectatorship.

As explained above, technogenesis is the process whereby technics create social and environmental paradigms through which the perceptual apparatus moves and adapts. The primary operation that engages this process is perception. Perception is the dual-directional action where the apparatus performs feedback and feedforward processes of information inputs and outputs. Working through the phenomenological arguments of Merleau-Ponty and Heidegger, Jason Farman (2012) explains that "our knowledge of the world and our place within the world depends on the feedback from our senses" (25). Our senses help us determine factors such as place, space, and time. For example, sight, sound, and touch combine to help us understand balance and extra-sensory operations such as proprioception. The senses also operate in continually shifting mini adaptations based on environmental factors. For example, one's sense of smell changes continually based on temporal exposure to an individual scent. Think of riding in a car down the highway and encountering a cattle processing stockyard. At first, the smell of fecal matter is overwhelming because it is a new and strong environmental presence. If you were to stay at that stockyard for an extended amount of time, however, your perception of smell would change as you would become accustomed to the offending odor's presence, and you would become less reactive to it. Your mind would tell the olfactory nerve that this smell is no longer offensive because it has become commonplace and it would be put into the background of its perceiving function, allowing it to sense other elements better. This is one way to consider how the perceptual apparatus adapts to environmental factors. Our bodily senses can adapt to social formations and technologies through heightened awareness or desensitization through our daily relations in what becomes our habitus.

In his discussion of participatory forms of performance, Gareth White (2019) shows how our habitus is also continually formed by social enculturation and re-enculturation through technics of technological mediation (140). White explains how our habitus can then be thought of as an enculturated combination of embodied cognition and affective sensing performed through the daily actions of a person's body. He states,

> The body that has learned how to interpret a social field, and is disposed to respond appropriately, has been shaped and habituated into an affective dynamic that allows it to function. A social situation is matched in the body by an appropriate affect, which drives a response-choosing moment, more conscious or less conscious as the situation allows. (148)

Our bodies are always orienting themselves toward the social constructs in which they are allowed to respond to through the perceptual process. Within social constructs of deep mediatization, they respond to many prevailing technics brought about by specific sets of technologies. Technologies are embedded in our environment physically, socially, symbolically, and culturally. Through this embedding, they become extensions of people's perceptual apparatus offering a form of hybrid sense of being for the mediatized spectator. When technics are considered in the relational mix as forces shaping and impacting one's environment, the mediatized spectator also forms a hybrid with more than just technological tools. Their perceptual apparatus becomes part of the environment from which perception emerges. This mediatized spectator is thus always in flux within its environment through a technogenetic form of sensing and perception utilizing the perceptual apparatus which can never be considered complete or even having a pre-given subjectivity as original subject divorced from its technological environment (Rosenberger and Verbeek 2015; Verbeek 2015).

As a human being's perceptual apparatus encounters technics, it learns to adapt to the way that technical environment performs. That constant act of adaptation forms the entire process of being-in-the-world, or put more simply, the entire experience of being human. As Farman (2012) states, "the senses connect us as being in the world through interaction" (26). Without the body/mind/environment amalgam that forms the perceptual apparatus, we might be nothing more than disembodied information or energy flows. The perceptual apparatus acts as the relational mediator for our entire understanding of experiential being. A mediatized perceptual apparatus is one that relies on connecting the process of technogenesis to the foundation of twenty-first century perception through mediatized social structures. This perceptual apparatus is unique from a supposed non-mediatized variety as the influence of technology and technics is a necessary agent for its operation. These are always present but not always accounted for. The terminology of a mediatized spectator is used to differentiate this form of subjecthood from more widely held conceptions of spectatorship that rely on ocular sight as a primary mode of information reception in favor of a more holistic and experiential mode of "taking in" through embodied acts of perception.

Within the fields of theatre and performance studies, this concept of the mediatized spectator also aligns with the posthuman subject whose construction is defined by its relation to technologies present within intermedial (Kattenbelt and Chappel 2006; Bay-Cheng et al 2010) or multimedia (Klich and Scheer 2012) performance. Returning to the discussion on Auslander and liveness earlier in the chapter, what separates the construct of the mediatized spectator and its perceptual apparatus from this posthuman model is the incorporation or reliance on technology—whether it be multi-, inter-, or transmedial—within or during the act of a performance event. A mediatized spectator is established prior to the event and thus can interact with(in) it whether it has technological involvement or not. While it may be useful to unpack a brief history of the incorporation of technology in and connected to performance, that is not the primary purpose of this project. Instead, I will highlight some of the primary theorization on the posthuman subject from scholars of intermediality in performance as an analogue for the mediatized subject who becomes mediatized spectator within experiential performance architectures. While the posthuman in performance and the mediatized spectator prior to performance have similarities, they are different in how they are discussed in performance scholarship. Performance scholarship has rarely taken the first steps of exploring the full impacts of mediatization as a techno-social force prior to analyzing the spectator in performance and instead has relied on the impacts of technologies within performance as the constructing force. By exploring these modes of performance, we can then work backward to better understand how a mediatized subject emerges from within its technological environment through the interactions of its perceptual apparatus. The scholars offered below discuss posthumanism and performance as a relationship that requires technological interfacing or technological use during/with the performance practices and objects analyzed. Mediatized spectators belong to a model that considers the relationship between people and technologies before the act of spectating.

In a comparable manner to Couldry and Hepp's discussion of media and society, Chris Salter (2010) and Steve Dixon (2007) establish a historical trajectory of the symbiotic relationship between media, technology, and human bodies by focusing on the incorporation of mediating technologies in performance. Salter discusses the technologically connected performer in *Entangled: Technology and the Transformation of Performance* and Dixon lays out a history of digital practices in performance in his book *Digital Performance*. Both books operate as critical accounts of the histories where media and performance merge prior to the social and phenomenological turn toward mediatization theory. Part of Salter's discussion centers on the historical influence of the digital paradigm, evolving out of the purely mechanical, and on the body in performance through the form of the cyborg. His analysis expands on the theoretical underpinnings of the posthuman body via Donna Haraway (1991[1985]), Judith Butler (1990, 1988), and Hayles (1999), among others, to give readers an understanding of the way in which bodies, both technological and corporeal, enter into constant interplay in performance. These cyborg bodies are hybrid entities often explored by performance artists "influenced by new electronic and computer-assisted technologies and informed by questions concerning new construction of subjectivity and identity" (249). Salter describes assemblages of humans and technology as "entanglement," which attempts to cover a broad scope of how performance and technology are interconnected and inextricably linked throughout history. Entanglement is a crucial way of thinking about the ecological perspective of interlinking networks of digitalization on which the mediatized spectator relies. Dixon goes further to place the theoretical basis of a posthuman subject in relationship to the slippage in subjectivity also discussed by postmodern and poststructuralist philosophers such as Derrida, Foucault, and Baudrillard. His paradigm emerges out of the discourse from these theorists' attempts to reconcile some of the uneasiness gained from the loss of grand narratives. Instead of loss, Dixon argues that the posthuman paradigm extolls a force "for cohesion, for meaning, for unity, for intimate cybernetic connections between the organic and the technological" (155). In Dixon's trajectory toward digitalization, the early work of Norbert Wiener on cybernetics lays the groundwork for Ihab Hassan's (1977) posthumanist subject revived by Hayles (1999). The focus of this trajectory was to define how people operate within a networked culture predicated on the operations of feedback cycles; the very inputs and outputs that define digital architectures and processes. Here, the human body gains a likeness to other digital machines and processes offering a model for a perceptual apparatus entangled with digital technologies. For Dixon, the posthuman is therefore a mediatized subjectivity that offers a possibility where "we are becoming media" itself when technologies become so ingrained within daily life that they disappear from our conscious frame of reality (153). What excludes Dixon's subject from my own is how he, like most others focusing on digitally augmented performance, focuses on the performer within digital and intermedial production practices without homing in on the subjectivity of the audience member who has been mediatized.

In a more concrete formulation, a posthuman subject is constructed and analyzed through the lens of transhumanism (a subset of the larger posthumanism) in works like Jennifer Parker-Starbuck's *Cyborg Theatre* (2011), which focuses on how technologies operate as extensions of the human body inside of multimedia performance. Parker-Starbuck's cyborg performance imagines "bodies on stage intertwined with the various technologies" present in the current era (8). A cyborgian performance is one that embraces a posthuman mode of storytelling but does so in a liberal humanist way of presentation and/or subjectivity. Her objects of study, such as Stelarc, are themselves cyborgian entities, which can be thought of as a performative mesh of the technological and the corporeal. Stelarc is both a transhuman and mediatized subject due to the purposeful technological manipulation of their perceptual apparatus as well as the invisible evolution of this apparatus through its interaction within its technological milieu. The mediatized spectator separates itself from the transhumanist construction of subjecthood because it emerges through the interaction with technics and technogenesis where human perception co-evolves, versus a physical upgrading of human potential through the material interface between software, wetware, and hardware. Mediatization impacts how one perceives and communicates in the world but does not require literally grafting technology onto the physical body of human beings as popularly expected of the cyborg formation.

Nuancing the transhumanist mode of application, Susan Broadhurst (2007) explains how Stelarc's performance practices relate to a mode of posthuman configuration that refers to "the transformation of bodies modified or *polluted* by technology" (87). Amending this mode of analysis, she explains that "human identities are mutated by the impact of various information technologies, which at the same time identify that impact. The post-human body is thus inscribed and reconfigured by its own mediatized and mediated narrative" (87). Through Hayles (1999) work, Broadhurst argues for thinking of technologically informed posthuman experience of reality to be one of interconnection, in the same way scholars of mediatization explain. The interconnection between technologies and embodiment destabilizes a conception of a technologized self and helps to reject a Cartesian mind/body duality negated through the application of embodied cognition and enactment. Broadhurst incorporates the technological (in this case the digital) as a crucial element with agency that connects humans to their environment and therefore operates as a distributed system of experience. This distributed system is what a mediatized model of experiential spectatorship operates through.

One of the first to explicitly connect a posthuman mode of performance to the act of spectatorship was Ralf Remshardt (2008, 2010). He explains, "In a posthuman performance paradigm, spectator and performer both relinquish their positionally determinate (dialectical) claims to presence and reconfigure themselves as dynamic, interdependent parts of an emergent system" (2010, 136). The analysis primarily lands on performance theory that considers technological mediatization as an integrated and relational part of the performance frame or what we commonly consider intermedial performance. He discusses the posthuman as an object in performance but restricts this subjective body to one that interacts primarily through digital remediation during the event, limiting the scope of mediatized spectatorship narrowly. Remshardt is important for more fully considering how performance theory must accept the concept of a posthuman and subsequently a mediatized turn; one which questions some of the foundational assumptions of performance studies. He asks us to reconsider the centrality of human embodiment and consciousness as the basis for performance analysis, and thus expanding possibilities for more relational analysis between dynamic objects, bodies (technological and corporeal), and informational systems (digital, performative, computational, communicative). It is this reconsidering that allows us to better understand how the audiences of contemporary performances operate through a mediatized subjectivity, one that is "dislocated and distributed" or always in flux and unfixed (2010, 137). This formulation of the body in flux matches well with the construction of the mediatized spectator and their perceptual apparatus which is always in relation with the technological forces embedded within mediatization and waiting to be shaped by the prevailing technics of its socio-cultural environment.

Matthew Causey (2006) takes up the socio-cultural environment of mediatization by discussing posthumanism in relationship to digital culture in ways similar to the above, but his argument is more explicit in how it claims that the prevailing theories relating to digital culture and performance were inept to adequately address a posthuman condition in performance. His claim is based on a belief that performance studies has an inherent bias toward humanization and he argues the field also over-relies on the human body as an ontological requirement of performance as an object of study. He states, "What mediated technologies afford performance theory is the opportunity to think against the grain of traditional performance ontology" (51). He introduces the terminology *postorganic* as a way of redistributing the human body-centered ontology of liveness necessary when doing any critical reflection on the digital and the virtual. He uses this terminology to "indicate the extensions and challenges to our bodies and selves brought on by the advances of new technologies" (53). Postorganic is a non-anthropocentric posture used to look at the influence of digitalization on human culture. Causey's theorization on these subjectivities after the impact of digital culture allows new discourse on the nature of spectatorship within domains of digital and/or mediatized performance. His work on the *postdigital* condition aligns well with the argument of techno-augmented posthuman sociality and is becoming a considerable influence on future scholarship on performance and subjectivity. Causey introduces the postdigital condition to develop a mode of thinking beyond binaries of digital and analog or live and mediated. The *post* does not indicate after or beyond in a linear sense. Causey (2016) explains:

My own model of the postdigital [...] considers the situation as such of a postdigital culture to be that of a social system fully familiarized and embedded in electronic communications and virtual representations, wherein the biological and the mechanical, the virtual and the real, and the organic and the inorganic approach indistinction [...] The prefix of *post-* linked to the internet, digital, inter-medial, and even the human is, in fact, a recognition of the overdetermined relations, circulations, and exchanges of those phenomena within the current condition—not an endpoint, but a recognition of the many flows and distributions. (432)

Causey's posthuman is an entity that exists in a postdigital technic that allows for a radical rethinking of the relationships between technology, the human body and mind, and the techno-social environments it operates with(in).

In a similar shift to Broadhurst, Rosemary Klich and Edward Scheer (2012) rely on the informatic technologized embodiment offered in Hayles posthumanism (189). They utilize Hayles' theorization to explain how the interface between the perceptual apparatus of the posthuman "being" enmeshed within systems of virtuality—one of the primary technics discussed later in this volume—offers a posthuman condition through which they analyze the works of Stelarc and the performance group Granular Synthesis. This analysis focuses on the materiality of information and media flows as they interact and shape the human body through "embodied perception" (180). They describe embodied perception as a "feature of multimedia performance" and utilize the case study by Granular Synthesis to show how a multimedia work impacts the spectator through material affects of its digital performativity, creating a form of posthuman subjectivity and embodiment (180). Within this mode of performance and interactivity, the spectator's perceptual apparatus enters into an exchange of sensed and performed information that replicates a posthuman perspective on embodiment. The crux of their analysis is to argue that "while most multimedia performance does not really merge man with machine, as a genre it problematizes perceptions of the separateness of bodies and media" (188). Klich and Scheer conclude their book by further focusing on the posthuman condition as an analog to the spectator's experience of reception in multimedia performance.

> This reception is an embodied reception that approaches the condition of posthuman embodiment: part actual/part virtual, part material/part information due to the positioning of the spectator's body between these domains. This experience of the virtual, for the spectator as for the performer, problematizes the parameters of their own sense of "presence". By presenting these two entities alongside one another multimedia theatre-makers force a comparison of the different ways in which participants connect with the material and the virtual. Not merging man with machine, but problematizing perceptions regarding the body and enabling the experience of new forms of embodiment and subjectivity. (205)

Their work aligns neatly with the concept of the mediatized spectator, but what is different in their deft analysis is a reliance on the mediating forces of technology or "technological intervention" within their objects of analysis (208). Their focus is on reading the performance itself, specifically examples of multimedia performance as posthuman which allows one to consider the totality of the event as well as the material conditions it enables within the exchange between performer, spectator, and spectacle. They explain this creates a "multimedial system" where the body is both "open to the work and affected by the work" (205) offering an architecture based on relational exchange. I intend to go further by expanding out beyond only examples of multimedia performance to show how their posthuman condition offers a mediatized spectator ready to encounter all forms of experiential performance, mediated or not.

Each of the above authors marks a beginning for a larger conversation concerning the way in which theatre and performance studies scholars might think about the way a posthuman mode of performance theory might help better situate discussions of the way the human and its perceptual apparatus change notions of spectatorship. Many of these authors introduce posthumanism through a thoroughly philosophical approach that attempts to work through the venues of deconstruction to destabilize the autonomous agency of the human subject who either serves as the central agent in either the making or the watching of a performance event. My framework for mediatized spectatorship takes up these challenges to a human-centered performance ontology to put into relation technological objects, technological processes, media, human perception, agency, and affect via the overarching social process of mediatization. The hope is that by both thinking in a relational manner about all these possible agents, we can extend the field of audience research to rethink the very ontology of contemporary spectators and subsequently the way they operate within various architectures of experiential exchange.

Note

1 Late-stage Millennials born 1988–1997.

Bibliography

Auslander, Phillip. 2021. *Liveness: Performance in a Mediatized Culture*. 3rd ed. Routledge.

Barad, Karen. 2003. "Posthumanist Performativity: Toward an Understanding of How Matter Comes to Matter." *Signs: Journal of Women in Culture and Society* 28, no 3: 801–31.

Bay-Cheng, Sarah, Chiel Kattenbelt, Andy Lavender, and Robin Nelson. 2010. *Mapping Intermediality in Performance*. Amsterdam University Press.

Broadhurst, Susan. 2007. *Digital Practices: Aesthetic and Neuroaesthetic Approaches to Performance and Technology*. Palgrave.

Butler, Judith. 1988. "Performative Acts and Gender Constitution: An Essay in Phenomenology and Feminist Theory." *Theatre Journal* 40, no 4: 519–31.

Butler, Judith. 1990. *Gender Trouble: Feminism and the Subversion of Identity*. Routledge.

Carr, Nicholas G. 2010. *The Shallows: What the Internet Is Doing to Our Brains*. W.W. Norton & Company.

Causey, Matthew. 2006. *Theatre and Performance in Digital Culture: From Simulation to Embeddedness*. Routledge.

Causey, Matthew. 2016. "Postdigital Performance." *Theatre Journal* 68, no 3: 427–41.

Chandler, Daniel, and Rod Munday. 2011a. "Perception." In *The Oxford Dictionary of Media and Communication*. Oxford University Press.

Chandler, Daniel, and Rod Munday. 2011b. "Perceptual Codes." In *The Oxford Dictionary of Media and Communication*. Oxford University Press.

Couldry, Nick, and Andreas Hepp. 2017. *The Mediated Construction of Reality*. Polity Press.

Damasio, Antonio. 2005. *Descartes' Error: Emotion, Reason, and the Human Brain*. Penguin Books.

Damasio, Antonio. 2010. *Self Comes to Mind: Constructing the Conscious Brain*. Pantheon.

Dixon, Steve. 2007. *Digital Performance: A History of New Media in Theater, Dance, Performance Art, and Installation*. MIT Press.

Farman, Jason. 2012. *Mobile Interface Theory: Embodied Space and Locative Media*. 1st ed. Routledge.

Farman, Jason. 2021. *Mobile Interface Theory: Embodied Space and Locative Media*. 2nd ed. Routledge.

Hepp, Andreas. 2013. *Cultures of Mediatization*. Polity Press.

Hepp, Andreas. 2019. *Deep Mediatization*. Polity Press.

Hansen, Mark B. N. 2004. *New Philosophy for New Media*. MIT Press.

Hansen, Mark B. N. 2006. *Bodies in Code: Interfaces with Digital Media*. Routledge.

Haraway, Donna. 1991. "A Cyborg Manifesto: Science, Technology, and Socialist-Feminism in the Late Twentieth Century." 149–181. In *Simians, Cyborgs and Women: The Reinvention of Nature*. Routledge.

Harvie, Jen. 2013. *Fairplay: Art Performance and Neoliberalism*. Palgrave.

Hassan, Ihab. 1977. "Prometheus as Performer: Toward a Posthumanist Culture?" *The Georgia Review* 31, no 4: 830–50.

Hayles, N. Katherine. 1999. *How We Became Posthuman: Virtual Bodies in Cybernetics, Literature, and Informatics*. University of Chicago Press.

Hayles, N. Katherine. 2012. *How We Think: Digital Media and Contemporary Technogenesis*. University of Chicago Press.

Hayles, N. Katherine. 2014. "Cognition Everywhere: The Rise of the Cognigtive Nonconscious and the Costs of Consciousness." *New Literary History* 45, no 2: 199–20.

Hjarvard, Stig. 2013. *The Mediatization of Culture and Society*. Routledge.

Iliadis, Andrew. 2015. "Two Examples of Concretization." *Platform: Journal of Media and Communication* 6: 86–95.

Kattenbelt, Chiel, and Fredda Chappel. 2006. *Intermediality in Theatre and Performance*. Rodopi.

Klich, Rosemary, and Edward Scheer. 2012. *Multimedia Performance*. Routledge.

Lavender, Andy. 2016. *Performance in the Twenty-First Century: Theatres of Engagement*. Routledge.

Lewis, William W. 2018. "Approaches to 'Audience Centered' Performance: Designing Interaction for the iGeneration." 9–26. In *New Directions in Teaching Theatre Arts*, edited by Anne Fliotsos, and Gail Medford. Palgrave.

Massumi, Brian. 2002. *Parables for the Virtual: Movement, Affect, Sensation*. Duke University Press.

McCutcheon, Jade Rosina, and Barbara Sellers-Young. 2013. *Embodied Consciousness: Performance Technologies*. Palgrave.

McLuhan, Marshall. 1964. *Understanding Media: The Extensions of Man*. Routledge.

Oddey, Alison, and Christina White. 2009. *Modes of Spectating*. Intellect Books.

Parker-Starbuck, Jennifer. 2011. *Cyborg Theatre: Corporeal/Technological Intersections in Multimedia Performance*. Palgrave.

Potzch, Holger, and N. Katherine Hayles. 2014. "FCJ-172 Posthumanism, Technogenesis, and Digital Technologies: A Conversation With N. Katherine Hayles." *The Fibreculture Journal* 23: 95–107.

Remshardt, Ralf. 2008. "Beyond Performance Studies: Mediated Performance and the Posthuman." *Cultura, Lenguaje Y Reprsentacion/Culture, Language and Representation* 6: 47–64.

Remshardt, Ralf. 2010. "Posthumanism." In *Mapping Intermediality in Performance*, edited by Sarah, Bay-Cheng, Chiel Kattenbelt, Andy Lavender, and Robin Nelson. Amsterdam University Press.

Roberts, Donald F, Ulla G. Goehr, and Victoria Rideout. 2005. "Generation M: Media in the Lives of 8-18 Year Olds." *Kaiser Family Foundation.*

Rosen, Larry. 2010. *Rewired: Understanding the iGeneration and the Way They Learn.* Palgrave.

Rosen, Larry. 2011. "Teaching the iGeneration." *Educational Leadership* 1: 10–16.

Rosenberger, Robert, and Peter-Paul Verbeek. 2015. *Postphenomenological Investigations: Essays on Human-Technology Relations.* Lexington Books.

Salter, Chris. 2010. *Entangled: Technology and the Transformation of Performance.* MIT Press.

Stiegler, Bernard. 1998. *Technics and Time, 1: The Fault of Epimetheus.* Translated by Richard Beardsworth and George Collins. Stanford University Press.

Stiegler, Bernard. 2009. *Technics and Time, 2: Disorientation*, edited by Stephen, Barker. Stanford University Press.

Stiegler, Bernard. 2011. *Technics and Time, 3: Cinematic Time and the Question of Malaise*, edited by Stephen, Barker. Stanford University Press.

Twenge, Jean M. 2017. *iGen: Why Today's Super-Connected Kids Are Growing Up Less Rebellious, More Tolerant, Less Happy — and Completely Unprepared for Adulthood*/*and What That Means for the Rest of Us.* ATRIA Books.

Varela, Francisco J, Eleanor Rosch, and Evan Thompson. 1991. *The Embodied Mind: Cognitive Science and Human Experience.* MIT Press.

Verbeek, Perter-Paul. 2015. "Beyond Interaction: A Short Introduction to Mediation Theory." *Interactions* 22, no 3: 26–31.

White, Gareth. 2019. "The Audience in Intermedial Performance." 136–158. In *Intermedial Theatre: Principles and Practice*, edited by Mark Crossley. Red Globe Press.

2 The Architecture of Immersion

2.0 Virtuality, Simulation, and Technologies of Immersion

Think about the world around you. How do you know what is real? What makes something real? Is the materiality of the thing you touch, see, sense what that makes it a real object to you? What about this book you are reading? Is it an physical object that you are holding in your hands, with pages you can turn as you move through the concepts and questions you are currently reading? How is this same or potentially different if you are reading it on a tablet or even more so on a computer screen? In each of these instances, virtuality and actuality are constantly in tension. With the computer and tablet versions of this reading, it might seem easier to understand the virtual over the actual. The "page" you see before you is a digital representation of the material manifestation of paper with ink forming words. *Because it is virtual, its intrinsic properties are mutable and in flux.* See, I just changed the style of the text with the flip of the switch, much like you can change the color of the page to address your reading habits. The computer screen is simply a technological tool used to virtualize a material form of symbolic coding using electronic signals. The light emitting from your screen is a form of materiality produced by activating pixels through electrical impulses. Does this materiality make the reality contained more or less real? With the tablet, you might gain a simulation of the tactile quality of book reading when given the ability to swipe as a simulated form of page turning. Does that make it even more real? It operates largely in the same manner. This must mean that the book is the closest thing to real, right? It is an object you physically manipulate, and its pages are the material through which you are given entry into a virtual world of the mind. Even though the words have tangible form in the book, they are still a tool of virtuality. The words printed here are simply symbols; virtual reifications of what I think, see, and interpret. These words are created to immerse you within the story of my mind, my interpretation, and my perception of the world. A story defined by my own sense of what is real. So, if this is all true then what is real? What is virtual? How do we locate the dividing line between the two?

This chapter focuses on immersion and its relationship to the ongoing influence of the technic *virtuality*. The effect of virtuality is like a fever dream;

DOI: 10.4324/9781003398813-3

something so detailed and engaging that when one is abruptly awoken, they believe they are still dreaming. They are immersed in this other reality to the point that it becomes hard to distinguish the difference between dream/imagination and the actual reality of daily life. The question I explore is: What happens to people when actuality becomes so imbued with virtuality that it becomes difficult to determine which is which? I argue these mediatized spectators become figuratively immersed in another reality to a point that their affective system of perception unconsciously seeks to escape the real of actuality. As this condition becomes pervasive, more objects of cultural production and artistic representation attempt to harness the power of virtuality by making themselves more immersive. Through immersion, the mediatized spectator seeks to access a deeper and more real "Real" via the affective sensing properties of the perceptual apparatus.

The etymology of the term virtual emerged in late middle English from the Latin *virtualis,* which roughly translates to almost real or relating to (-alis) something virtuous or real (virtue). One of the earliest uses of the term defines virtual as "relating to essential, as opposed to physical or actual, existence," (c 1443, OED Online) relegating the virtual to the not-quite-real. In this early usage, essence is the philosophical/spiritual basis of beingness that brings one into being and forms much of the basis of metaphysics. The virtual is a counterpoint to the foundational concept of the Real (actuality) as a rationally explained construct through which people make sense of the material world. This making sense comes about through the operations of one's perceptual apparatus sensing and processing input from material and symbolic sources. The virtual as an analog to the essential has existed since humans could question their own beingness and becomes the technic virtuality when it is applied through tools used to make sense and shape reality. The essential informs the actual through virtuality and vice versa, creating a feedback loop of meaning-making and meaning to be made. Virtuality is therefore a technic that expresses itself through various tools and media but is also a mental figuration that exists outside these technological tools.

The concept of virtuality comes in a variety of flavors that suit individual ways of use and fields of study with two primary intertwined perspectives: That of the metaphysical/essential and that of the actual/technological. In the past century, virtuality became all-encompassing as digitalization developed into the prevailing wave of mediatization shaping societal relations. This follows N. Katherine Hayles' (1997) proclamation that virtuality had become a condition in late twentieth century societies which acts as both a "cultural perception" and a "mindset that finds its instantiation in an array of powerful technologies" (184). To best understand how virtuality operates as a technic, it is necessary to probe the places where the competing strands overlap and impact each other, or as Hayles describes, "The perception facilitates the development of technologies, and the technologies reinforce the perception" (184).

The first hinge of this chapter focuses on the technic of virtuality that technologies of simulation—virtual reality (VR), video games, surround sound for film and television, and theatre—harness to give the mediatized spectator a felt sense of belonging with(in) a virtual environment. Virtual environments are designed to give the spectator a perceived sense of being immersed with(in) the not-quite-real space of virtuality and offer the spectator a remove from the space of actuality through various media. Through the technic, people habitually traverse the line between the virtual and the actual. After time and deeper connection to virtuality, this technic instills an unconscious sense in people, where immersing oneself in the virtual can become so appealing that it can seem more real than the actual. Through aspects of digital immersion and simulation, cultural products amp up the experience of the virtual, making it more visceral and impactful until they replicate the actual. This feeling of impact is further amplified by cultural narratives supporting a desire to transcend the actual, just like dreams do. In response to these narrative and digital born realities, immersion as a form of exchange is then harnessed in representative media through heightened bodily affect as a way of piloting back those who have sought escape in the virtual.

After defining virtuality's impact on contemporary life and exploring scientific, philosophical, and artistic perspectives of its effect on peoples' perceptual apparatus, a socio-historical survey of simulation technologies that harness virtuality is explored. This prevailing technic became pervasive and affective within digitized society, bringing about an emphasis on immersivity within the experience economy. The focus is on technologies emerging in the late 1970s whose cultural fascination reached ubiquity shortly after the birth of the commercially available internet. Virtuality is primarily explored through VR tech, early video gaming, movies, and television, showing how it became commonplace by the end of the 1990s. Immersion is then explored through performance and media perspectives. By connecting virtuality as a social force to technologies which spectators become intertwined with via digitalization, it becomes possible to explain how the proliferation of immersive cultural aesthetics found in experiential media are contemporary artistic representations of virtuality through the exchange properties of immersion. Immersivity in these cultural products offers mediatized spectators the actualization of virtuality through the interaction of their perceptual apparatus with spaces, places, and narratives utilizing affective exchange.

The second hinge of this chapter offers Instances of spectatorial exchange with(in) immersive media to layout an argument for how the *Architecture of Immersion* primarily operates through bodily affect and a perceived sense of agency brought about by simulation and felt experience. This is necessary to pick apart the ways immersion is often casually applied as an overarching umbrella term for the many architectures of experiential spectatorship. Instead, I offer immersion as a foundational system of exchange that allows spectators to transport themselves into more than or not quite real (depending on your perspective) worlds to experience what it feels like to belong with(in) another reality. Immersion is therefore a process that works as a type of affective glue to hold spectators transfixed with(in) these simulation systems utilizing logics of virtuality. The goal is to bracket the experience of immersion away from other experiential forms by defining the ways in which a mediatized spectator interacts with virtuality through affective agency and exchange.

Virtuality as Technic: A Techno-Philosophical Perspective

When considering the impact of technogenesis on the perceptual apparatus of mediatized spectators, arguably there has been no greater influence than the concept and operation of virtuality. Virtuality sets the foundation of all the other technics discussed in the following chapters. Building off philosopher Brian Massumi (2002), Sarah Bay-Cheng (2010) argues that virtuality "occupies a crucial space between what is imagined and actualised, between potential and realization" (142). Virtuality as a technic performs the role of mediator between two "Reals": One that is seen and felt, which we will call actual/actuality, born out of material realities, and one where what is seen and felt is simulated through technological tools and discussed as virtual/ virtuality which offers an escape from the material realities that might limit human action. To begin, one must first see how virtuality has historically worked as a counterpoint to the concept of reality. Following a historical trajectory shows how this powerful technic has helped to develop the contemporary mediatized condition formed out of relations between people and their simulation technologies. This section begins by highlighting the importance of virtuality as a crucial technic that has grown as societies have advanced technologically and which has become embedded in these societies beginning near the end of the twentieth century.

Before jumping into the history, let's first establish some frames for how we might understand the term virtual within a techno-social frame. Philosopher of cyber-culture Pierre Lévy (2001) deconstructs the concept of the virtual into five areas of understanding: (1) *Common*; (2) *Philosophical*; (3) *Information Technology*; (4) *Information Systems*; and (5) *Narrow Technological* (56). His list ranks each meaning from weakest to strongest in terms of how they are perceived and concretely understood and focuses on digital technologies as the primary tools that help people traverse Reals. The *Common* meaning of the virtual is "something false, illusory, unreal, imaginary, or possible" (56). This follows the earliest understandings of virtuality defined through the language above. This mode of the virtual is that which cannot be made actual, as it pertains only to that beyond what we once can believe as a "true" reality. The *Philosophical* understanding of the virtual is something that exists only in its potential versus its actuality or "something that exists without being there" (56). Lévy's example is the tree that a seed will become versus the tree that already exists fully grown. Here lies the basis of the metaphysical understanding of virtuality, where perceptions and understandings of how the world exists allow one to believe that another reality has the potential to come into existence. It is not necessarily a binary between actuality and virtuality but rather a potentialization of virtuality becoming actuality. In *Information Technology*, the virtual is a set of possibilities calculable from the interactions between user input and a digital model/system. This includes the multitude of messages set by models such as "software for writing, hypertext systems, and interactive simulations" (56). This understanding marks the annexing of virtuality into technological realms where the philosophical gains actualization. Possibilities only arise through input from the user and are indexed alongside the digital coding used to translate the input into actualities. The coding is fixed and establishes a technical framing for these possibilities. For example, when typing on my keyboard, letters arrive on this screen/page. When the page was blank, the words inputted have the possibility of being *Hamlet*, a haiku, leetspeak, or simply gibberish. While typing, the tool translates my thoughts into a system of logical representation formed out of the letters emerging based on the keyboard key. Keystroke becomes signal, signal is translated into code, and code is represented as letter on the screen. Each letter creates a combination of words establishing a framework for reality based on language. Information technologies act as the actualizing mediator between action and system but are still largely intellectual due to the limitations of the digital model. Information technologies are those that help to make the virtual make sense logically. In *Information Systems,* a referent exists that represents a "message of space proximity" that the user has control over in some manner (56). Lévy lists video games, virtual realities, and networked role playing as examples. In each, an avatar exists as a link between the digital map-space and the corporeal world-space. The user moves this avatar through the digital space virtually through some form of technological controller. The user's input allows a positionality and subjectivity to emerge where the movement of the digital avatar simulates the viewpoint of the user but does not fully replicate the embodied perceptions of reality. In this sense, the user becomes immersed within the digital space of virtuality but primarily only as a mentally processed perception even though there is often physical manipulation through a controller. Even so, one could not actually feel what an avatar feels, but they could imagine that feeling based on the fidelity of the image/sound generation. In the most concrete concept of the virtual, *Narrow Technological*, there exists an "illusion of sensori-motor interaction with the computer model" (56). This might include the use of data gloves or haptic suites in VR simulation programs. Through technological augmentation of the person's body, corporeal interaction translates to virtual movement and response. As one moves through each understanding of the virtual, one can see how a shift from a mind-centered reality is supplanted by a technologically enhanced body-centered version. In each example, the level of actualization understood is based on immersivity. Lévy explains, as computer networks expand, they form the "informational universe of virtuality" and "the more they expand, the greater their power, storage capacity, and bandwidth, the greater number of virtual worlds and the more varied they become" (57). As the information universe becomes pervasive, it gains the ability to supplant the logics of virtuality based on mental immersion and relies on affective immersion through one's perceptual faculties. When Lévy established his schema in the late 1990s, the world had just begun to feel the impacts of the internet and digitalization, but the potential of entering virtual worlds had become a central concern engulfing all aspects of daily life.

Virtuality is explored using each meaning individually but ultimately lands on an amalgamation because of the way each is interlinked via contemporary digital media. In today's complex mediatized world, virtuality engulfs nearly all elements of daily life. As such, it is impossible to remove oneself from the condition of virtuality, as one is always immersed in its illusory operation. As will be shown in the following parts of this chapter, virtuality and the subjective positions it creates become a backdrop from which other forms of spectatorial exchange occur. Virtuality serves as a conduit through which digitalization operationalizes people's perceptual apparatus, offering both a deepening and resistance to its effects through experiential forms of spectatorship.

In the foundational stages of application, virtuality has more to do with an emergent property contained within the walls of the human mind/imagination where one can "transcend limitations or external determinations" (Foster 2009, 319) defined by actuality. Virtuality is therefore first a symbolic force born out of the desire to escape the confines of the actual and to enter into a real beyond the actual. This escape has historically surfaced through various narratives, philosophical discourses, and forms of artistic representation. Philosophical conceptualizations of virtuality go back to Plato, are complicated by Descartes, and reach their peak potential with the philosophical debates emerging during the turn toward poststructuralism and postmodernism. Philosophers such as Derrida, Deleuze, and Baudrillard offered nuanced and sometimes competing perspectives on virtuality as a psycho-social force defining many of the effects that decenter the logic of the human in relation to language, technologies, and everyday structures of material reality. Their work also informs many of the scholars of new media and cybercultures emerging in the later part of the twentieth century who begin to co-join the philosophical with the technological when the capacity for actualizing the virtual through digital technologies becomes more readily possible and available.

It is best to start with how reality has been understood historically within the frame of Western cultural systems.[1] A brief survey of some of the most significant philosophical perspectives that have foundational influence on virtuality as a technic, both in the philosophical and technological sense, is necessary. In Plato's *Allegory of the Cave,* the question of reality and representations of reality are explored to help define proper behavior in a structured society. Within the cave sits a chained person who can only see what is directly before them. What they see is shadows on the wall. These shadows are the only things they can perceive as reality because they have nothing else to compare them to. Little do they know that the shadows are being created by invisible "puppeteers" utilizing the light of the world outside the cave. The light represents the essence where true reality exists, and the puppets are the tools that artists use to capture these ideal/essential truths in material form. When the chained person escapes and encounters the essential, a binary emerges that compares the shadows (material representation) to the light (virtual essence). A second binary is also formed: Artistic interpretation versus ideal form. Within a political frame, this binary is used to blame artists as corrupting forces on the nature of ideal citizens. These binaries become philosophical precedents for how virtuality historically conditions one's perception of reality and existence. A question persists that drives people forward to seek out a truth. What is real: What one perceives or what one believes? Is reality defined by one's interior mental logic or one's external perceptual processing? Plato's binary formulation is among the basis for much of Western philosophy as it establishes the basic questions pertaining to existence and people's understanding of place within it.

During the Enlightenment, further questions of one's responsibility within the real world persist and lead René Descartes to question the very nature of reality as something that exists beyond the human mind. Here, he builds off the Platonic strand of philosophy to establish Cartesian duality that is based on a body-mind binary driving much of philosophical thought until the mid-twentieth century. Later, that dualism began to break apart partially due to the turn toward phenomenology which looked seriously at the subjective relationship between people, technologies, and experience. Descartes questions whether a person can know if the world they perceive is anything other than an illusion, a dream, or a magical affect conjured up by an evil demon (Chalmers 2022). In each example, the possible external forces that facilitate perceptions of the world around a person could all be a lie or a virtually constructed reality. To counter this issue of external reality, Descartes proclaimed that reality must exist because a person exists, but that existence is only made real when one thinks it into being. With this proclamation, the concept of "I think, therefore I am" gains traction. Under this logic, the body of the person becomes unnecessary in the construction of the world because what it perceives could be unreal. Reality then only exists for one who can think it into existence, and what they perceive through their body is something that can be deceived and is therefore not to be trusted as the basis of reality. Following this logic, anything that is beyond the base level of consciousness within the brain as a processor is subject to being made non-real or a virtual reality. What this new conditional forming of mind-body duality institutes is the potential for the virtual—or what is possible—to influence what can become actual. Virtuality as a social-technical force is embedded in the concept of the brain being the creator of reality.

The signal processing metaphor is also referred to as the brain in the jar (or vat) problem introduced by Hilary Putnam in his 1981 book *Reason,*

Truth, and History. By combining Descartes "cogito, ergo sum" with technological tools one can hypothesize the possibility of a brain detached from the human body and placed in an electrolyte liquid inside a container of some sort that is connected to a sophisticated computer sending electrical signals to the brain. With the brain still able to process electrical impulses (materiality), it could still develop a conscious construction of a real world independent from its body. The perceptual apparatus is reconfigured with the liquid in the vat representing the material substrate of the body with all its processing sensors. The brain as a signal processor metaphor is where much of science fiction relating to virtual realities is based and informs much of the contemporary concepts of virtuality. This is one of the foundations for *The Matrix* (1999), where human bodies still exist but are simply non-sensing power sources for machine societies. Future philosophers extended this train of skepticism to focus on how social forces can then be manipulated through technologies and media which then change the signals, scrambling how reality is perceived.

In the early part of the twentieth century, Martin Heidegger pushed back against the proposition that reality is based on virtual constructions of the mind and argued that reality is intersubjective and based on the material experiences of the individual interacting with the world around them. His basis of phenomenology focuses on events or phenomenon one experiences which are what define reality, not the virtual background of possibility that the mind might dream up on its own. Maurice Merleau-Ponty then extends Heidegger to focus on the senses of the body—gained through worldly interaction—where a mind gains the information it can process to define the world. Phenomenology is fundamentally a philosophy of embodiment which questions the reign of the Cartesian mind/body duality. Phenomenology might seem to then end the potential of virtuality as a technic, but Heidegger gives it new force by seriously examining the agentic capacity of technologies as an extension of what makes up human realities. He explains that a tool is only made tool like through its human use. And through its use, the tool also extends the potential of the human being. Through one's interactions with material actualities, like technologies, the individual gains their own intersubjective sense of being-in-the-world, their *dasein*. Under his formulation, the foundation of beingness is based on worldly interactions but that world is also formed through those interactions. Neither exists prior to the other as they are both made (through the formation of meaning) out of the co-construction of the other. Heidegger's theory extends the importance of how virtuality informs our human condition through his emphasis on technologies and embodiment. He extends the questioning of relations between the possible and the real and opens the door to an even stronger set of shaping tools that come when the digital age is born which parallels the rise of postmodernity.

The postmodern turn was fundamentally a socio-cultural tipping point where the ontological and epistemological questions found in philosophy begin to bleed into each other, creating a cross-pollination among a variety of fields. Of particular importance was how meaning was made through various technologies of representation or media. As explained previously, all media (language, writing, print, performance, etc.) are technologies that shape human potential and co-evolve within human social processes. During the postmodern turn, media and technologies began to take a more significant role as they have a deep connection to communication and representation of realities and virtualities. They also help to defuse the emphasis on the binary relationship between virtuality and actuality. Gilles Deleuze's work looks back to phenomenology and critiques its rigidity over the course of his career, making important contributions to the argument that there is no such stable ontology of human beingness, but only an unending process of becoming. Being is replaced by becoming which then allows the reformulation of real and possible to be replaced by a fluid relationality between the virtual and actual. In this capacity, the Real forms through the interplay between the virtual and the actual. This maps on to what Deleuze and Guattari (1987) describe as the rhizome in their work *A Thousand Plateaus*. A rhizomatic perspective on reality operates from an assumption that beingness is based on a non-directional hybridity and non-structural structuring. This subject always in-process situates itself within an unlimited network of inter-related agents. They explain the rhizome as a system with,

> no beginning or end, but always a middle, from which it grows and which it overspills. Unlike a structure, which is defined by set points and positions, with binary relations between the points and biunivocal relationships between positions, the rhizome is made only of lines: lines of segmentarity and stratification as its dimensions, and the line of flight or deterritorialization as the maximum dimension after which the multiplicity undergoes metamorphosis. (21)

Using the schema of the rhizome, the virtual is an infinite set of potentialities always accessible through the actual which materializes or is made concrete, *at least for a brief moment*. That moment then introduces an exponential number of new potentialities. Virtuality now becomes the always already accessible to be made actual through various technical interactors. Actuality allows the virtualization of new potentials, leading to a never-ending set of relations which questions all foundational theories of ontological stability within society and culture.

With the sociological scrambling of postmodernity as a backdrop, no such thing as a stable reality enters the picture. Jean Baudrillard (1995a) described reality as a simulation in his foundational writing on technology and postmodernity, *Simulacra and Simulation*. Simulation becomes his replacement terminology for virtuality, and all technologies of simulation likewise are tools for representing the virtual. He deconstructs the nature of late 20th century reality as one without an origin, as simply a "hyperreal" where the creation of a social reality is made by the duplication of imagined realities. This harkens back to the anxiety of creative representation in Plato's cave allegory. He argues that actual reality is indistinguishable from any virtual version because society has progressed beyond any definable original and stable real which has instead become hyperreal. In Baudrillard's hyperreal version of society, it is only virtual expressions of truth (film, art, media) that can even come close to approximating the essence of reality because they do so via a distillation or exaggerated effect, essentially making the actual too mundane to live up to the expectations of the ideal. Nothing replicates the feeling of this condition better than *The Matrix*. Those traveling through the tunnels above Zion must compare their lives to the ultimate fantasy of the simulated reality inside the matrix. For instance, when Neo awakens from his "dream" he is blinded by the light of the ruined society, with Morpheus declaring "Welcome to the real world" (Wachowski and Wachowski 1999). Neo is introduced to his previous subjective captivity, harkening back to Descartes's evil demon scenario, this time the demon being replaced by sentient machines. Later in the film, one of Neo's shipmates Cypher makes a deal with the system in the guise of Mr. Smith to be plugged back into the matrix because reality is less appealing to life in the simulation. While cutting into a simulated steak, he explains, "You know, I know this steak doesn't exist. I know when I put it in my mouth, the matrix is telling my brain it is juicy and delicious. After nine years, you know what I realize? Ignorance is bliss." This epitomizes how Baudrillard saw contemporary society. The hyperreal had become the only reality accessible as true. In the 1990s, this idea of simulation, as a "stable" and preferred reality, took further hold of the cultural milieu of technologically advanced societies. As digital technologies began to replace analog tech, simulation became a stand-in for the concept of virtuality. More computing power allowed an exponential expansion of potentialities within digital domains. Virtuality as a technic then gained a new power in the material processes of the digital as they began to supplant the imaginary forces of the virtual. These technologies could produce simulated realities at an alarming pace and with such force to overcome the virtuality/actuality binary, leading to a paradigm where actuality is subsumed by virtuality and thus becomes its own actuality.

Baudrillard continued to develop his theoretical relationship between simulation and reality by explaining how the society of the spectacle had fully engulfed the perceptual apparatus of spectators, turning actuality into a less authentic reality. Virtual realities were the place where the potentials of human imagination could be made real, or at least make people believe they were real. Using Operation Desert Storm as an example, he argues that war is only made real in its intricately manipulated filmic editing and televisual re-broadcast (1995b, 2014). Reality only exists through its technical mediation. It is no surprise that the 1990s also saw the rise of "reality television" where the carefully crafted and edited antics of "real" people were displayed as a way of heightening and theatricalizing the mundane. Baudrillard intones this phenomenon as the beginning of the end of the spectator: "There is no separation any longer, no empty space, no absence: you enter the screen and the visual image unhindered. You enter your own life as you would walk on to a screen" (2014, 193). Simulation and virtuality become the cultural way of life as depicted by the televisual and other forms of mediated communication under digitalization. When virtuality engulfs actuality and performing replaces *being*, everyone becomes immersed in an aesthetic of the spectacle as mediatized spectators. One aspect that Baudrillard seems to be less concerned with, however, is the relationship between the human body and how it informs and is informed by this new paradigm.

On the heels of Baudrillard, N. Katherine Hayles (1997) defines the *condition of virtuality* as a definitive end to the demarcation between the real and the actual, using the terms virtuality and materiality. Due to its connective properties, and its remediation of the operations of the televisual, the internet began to bring virtuality into nearly every connected home, imbuing humans with a technogenetic predisposition toward technological immersion where new forms of reality are accessed through digital products. She argues that shortly after the cybernetic wave of the 1940s, binary code allowed virtuality to seep further into technological domains via computer-generated realities. Virtuality moved into the realm of digitality, and with the advent of better processing power and graphic-based interfaces, a basic code for digital beingness inside virtual domains became possible. Combining the focus on embodiment from phenomenology with the processual focus of Deleuzian ontology, Hayles offers the construction of a virtual subject that is formed through their interaction with simulation technologies. The condition of virtuality she offers extends forms of perception and interaction, opening up the sphere of influence to all aspects of digitalization. This condition is pervasive and changes everyday life, not just life with machines. Of specific concern in her early work is how digital textualities and graphic worlds change the relationship between virtual (digital) and actual spaces. The human who has historically been situated within actual environments with material objects now increasingly interacts with simulations of those "actualities" in digital space. What would have previously been the domain of the mind gains material qualities based on informational patterns of code. For example, think of the visual representation of the matrix in the movie by the same name. To viewers of the film—and at first Neo as well—it is simply a cascading scroll of electronically lit symbols on a computer screen, but for those operators who have been habituated to its language/code they can see and perceive the entirety of the virtual world contained within the program, thus creating a new form of posthuman or cyborgian construct via technologically augmented perception. Hayles (1997) offers a warning for those that see information (digital code) as simply a product of technology without consequences for both subjectivity and embodiment: "Virtuality is not about living in an immaterial realm of information, but about the cultural perception that material objects are interpenetrated with informational patterns" (204).

Hayles also challenges the idea that the human body is based purely on material properties because it becomes enmeshed with digital systems that support the technic of virtuality. The body extends into the informational realm as a form of mediatized proprioception, allowing the virtual to have material properties and consequences. When the condition of virtuality becomes the everyday, informational realms of the digital offer a new formulation of beingness that is based on a technologically driven phenomenology. Hayles' subject/subjectivity born out of virtuality is offered as posthuman. In her most influential project *How We Became Posthuman* (Hayles 1999), Hayles describes the construction of the posthuman subject as "an amalgam, a collection of heterogeneous components, a material-informational entity whose boundaries undergo continuous construction and reconstruction" (3). She introduces the notion that people might have always been posthuman by applying the logic of technogenesis to the historical formation of human beingness. Hayles' focus on historical relationship between humans and technologies also maps onto the formulation of historical waves of mediatization.

Following this historical trajectory, the seeds of the posthuman or original mediatized subject were planted at the very beginning of people's use of technological tools. During the Industrial Revolution, roots began to take hold via mechanization and were further cemented when electrical signals began to expand the reach of mediated information. With the introduction of cybernetics in the 1940s, the fully realized conception of a posthuman subjectivity emerged and introduced a fusing of the philosophical and technological aspects of virtuality. Cybernetics developed into a form of empirical study that considers a human being as a form of information or informational flow of consciousness that could possibly be removed from its embodied flesh and transplanted into a machine, like the brain in the jar hypothesis. The difference was that with the advance of cybernetic technologies the potential for artificial or machine intelligence was made possible, at least in theory.

The cognitive machine, as a replacement for the biological human being, is the first step in the formal creation of the technological cyborg: A liminal creature whose being is determined by the overlap between the machine and the human. Donna Haraway's (1991[1985]) concept of the cyborg works as a metaphor for the "historical transformation" of humans beyond a mind and body dualism toward a new humanity based on the merging of "imagination and material reality" (118). The cyborg is therefore a "creature able to bridge the gap between the real and representation, between social reality and fiction" (Giannachi 2004, 46). This is a crucial step in the merging of virtuality with embodiment through technological means. By thinking through this cyborgian figure's potential, Haraway offers a new way of approaching the techno-scientific understanding of the human, to create a slippery liminal figure that exudes political power that can look past identity politics as such due to an emphasis on deconstructing the notion of identity based on an essentialist foundation. This figure also became an ideal stand-in for the new body politic emerging out of an era informed by cyberspace and cyberpunk alternatives to the future.

Due to these increasingly ubiquitous connections to digital interfaces, a fixed and stable understanding of the individual self, and the world that contains this self, is increasingly difficult to maintain. This also means that a clear divide between actual and virtual worlds is even more difficult to establish. As digitalization became embedded, technological societies became unconsciously aware that the gap between virtuality and actuality was disappearing because of increasing interactivity with digital material. In the late 1990s, the growing influence of computer-generated patterns of information began fueling a paradigm shift in the way human beings perceived their sense of being and place in the world. Hayles (1999) states, "When I say virtuality is a cultural perception, I do not mean that it is merely a psychological phenomenon. It is instantiated in an array of powerful technologies. The perception of virtuality facilitates the development of virtual technologies, and the technologies reinforce the perception" (14). Technologies such as computer applications, the internet, and video games that use graphical interfaces—without the need for a user to understand computer programming language—operate through modes of visual and affective fidelity that can reasonably convince users that the virtual worlds contained within their domains are real enough to exist as parallel dimensions; ones gaining equality with that of actuality. By gaining an ontological equivalence to actuality, virtuality could imprint on the perceptual apparatus of mediatized spectators a sense of dual Reals: One based on the "functionalities of a computational universe" (15) and the other based on the operations of affective sensing through embodied perception.

As the computational universe becomes more pervasive, it begins to supplant the epistemologies of the corporeal and subsumes both the bodies and minds of human beings to the point that virtuality begins to become "more essential than material form" (17). At this historical point, the prevailing way of life is based on the condition of virtuality. Hayles' early work on virtuality came well before the mass adoption of the cell phone, not to mention the smartphone, tablet, or commercially available VR systems that are commonplace today. She marked a fundamental shift in how scholars contend with the virtual/actual dichotomy moving forward and set the stage for advanced theorization on the relationship of media to society, culture, and aesthetics. In her later work referenced in Chapter 5, she doubles down on the technogenetic influence of digital tech as something that not only extends the body but also rewires our cognitive abilities.

The theories above have been hugely influential for theatre and performance studies scholars working in the areas of digital, intermedial, and multimedia performance. Matthew Causey (2006) is a standout due to his focus on creating deep connections between the philosophical and the technological strands of thought on virtuality and connecting them to performance in the 21st century. He explains that the most recent rush of theory on the technological strand of virtuality began with the introduction of cyberculture in the 1980s:

> The virtual, issuing forth from the televisual and the cyberspatial, is the technology by which we know today. At the core of the experience of the virtual is an avoidance of those technologies of representation that excavate the double of the theatre and the mask, thereby denying the corporeal nature of the body and the mind. (97)

It was in the early 1980s that both theorists and hobbyists began to explore the possibilities of the virtual realm as depicted in the representative cyber worlds of computer-generated fiction, games, and chat rooms. In these digital domains, users began to experiment with new modes of identity creation and imaginary world building. During this decade, the foundations of the technology that would become known as VR started to enthrall the imaginations of a culture quickly engulfed with all things digital and cyber. I remember as a pre-teen in the late 1980s watching episodes of *Star Trek: The Next Generation* where the advanced technology of virtual reality was imagined in the form of the holodeck. This VR simulator allowed members of the crew to take "vacations" and engage in games and fantasies. In the many episodes that used the technology, Data, Ryker, and Picard (among others) enter the digitally rendered virtual world of Alfred Conan Doyle's *Sherlock Holmes* fiction. In the holodeck, virtual realities become more than virtual as they are fully interactive in a corporeal manner. The holodeck represents an epistemological shift in how spectators perceive the binary between the virtual and the actual. This shift is most fully explored in the episode *Ship in a Bottle* (Singer 1993), in which the elusive Dr. Moriarty wills himself into the actual reality of the Enterprise. The holodeck serves as a metaphor for the illusionary world of our consciousness as the technology for perceived actuality. In the same episode, Picard ponders whether their reality aboard the space vessel could itself simply be a program. Like the fantasies in the holodeck, could their entire existence only be an intricate set of simulations played out for another cosmic spectator? This question is mirrored by Causey (2006) when he warns that "the era of virtuality (a problem of illusion and representation), the age of simulation (e.g., the televisual representations of Gulf War I), has given way to a more troubling model of embeddedness, a problem of materiality and embodiment" (151). He is concerned with how virtuality and simulation have engulfed the social consciousness of everyone, and this potentially endangers the future of live performance forms as uniquely separate from their mediated counterparts.

Arguing that the warning had come true, Causey (2016) later marked a new paradigm created and nurtured by an encompassing of actuality by virtuality; undoubtedly extending Hayles' condition of virtuality, he names this paradigm the "postdigital condition" (428). In this structure of virtuality

enmeshed with(in) actuality, reality is no longer explainable via a divide between the virtual and the actual. He explains:

> The reality of the virtual is perhaps the most complex of these articu-
> lated modalities of the digital, but it indicates that the binaries of the
> biological and the virtual, the organic and the inorganic, the machine
> and the flesh, and specifically the virtual and the real are no longer
> useful in conceptualizing and performing within a postdigital culture.
> Those categories are recognizable and still inform reason and logic, but
> are increasingly active in *zones of indistinction* as indiscernible phe-
> nomena. (434)

The two zones are forever intertwined, creating a dynamic tension for
those that encounter the spaces where the two meet, as if there is a magnetic
field between the two polar extremes. A postdigital subject is both a mediatized and embodied traveler sitting in a liminal field, between, but also,
within the two, attempting to locate its physical presence in the actual world
through affective sensing while its mind tries to navigate the multitude of digital spaces created in the computational universe of virtuality. The postdigital
condition forces us to "think like a machine, digitally, or risk obsolescence"
(440). Doing so alters a person's perceptual apparatus and moves them
"toward a more fundamental encounter, an even more unsettling event: seeing oneself as no longer just human, but in a position as posthuman, becoming machine and thinking digitally" (440). This way of becoming is at the
heart of the current logic of virtuality.

The survey of these various perspectives on virtuality is just the tip of the
iceberg but is intended to show a trajectory of how the technic has evolved
overtime and conceptually progressed from something abstract and beyond
true human comprehension into a mode of accessibility via technological
means. The philosophical perspective focuses most on the potential for people to better explain and understand the force of imagination and the unknowable that lives beyond the actual, while the technological attempts to
identify the material elements of one's world that allows one to access that
unknowable. By merging the two strands together, theatre and performance
scholars can examine how experiential performance works as another form
of virtuality, where transcending the material and entering the virtual becomes possible. In an era where virtuality becomes a way of being in the
world, harnessing its capacities becomes crucial. Understanding how the material influences of technology channel virtuality is of particular concern and
is the focus of many of the cultural products discussed in the next section.
The Matrix has been referenced multiple times throughout as it is a prime
example of an exploration of the philosophical and technological. Neo is a
stand-in for the mediatized spectator, one who desires to gain power over
the mediating nature of virtuality by becoming one with the machine while

still preserving their humanity outside the matrix. Doing so allows them to enter into a place of relation between the two and gives them the ability to shape their own experience. Like Neo, a mediatized spectator is one who transcends the limits of actuality while still maintaining engagement with its physicality and materiality. So, will you take the red or the blue pill?

Immersed in the Virtual: Simulation Technologies and the Promise of Transcending "Reality"

Virtuality is not only a philosophical technic. It is embedded in the technologies of representation of the individual era. In antiquity and through the Middle Ages, it was primarily a form of meaning-making through oral storytelling and human contemplation. Stories helped to connect the human imagination to material realities through an ability to escape into fiction and narration. During the Age of Enlightenment, virtuality became increasingly technical as stories could travel with the printed word, offering a plethora of new virtual spaces for the imagination to explore. With the number of narrative worlds increasing exponentially, readers could metaphorically transport into the virtual space of the imagination, increasing the social power of virtuality. Janet Murray (2017) invokes Cervantes' *Don Quixote* when stating, "The dangerous power of books [is] to create a world that 'is more real than reality'" (124). The fictive world is expressed in two spaces: The actual printed page of the book and the virtual world enacted in the reader's mind. The reader becomes immersed via contemplation, imagination, and deep thought in the virtual worlds invoked by the text. Around this same time, the expansion of cartography via mechanical sources offered a new artistic medium for capturing the complexity of the world via pictorial representation. Maps of far-away lands and seas imprinted virtual impressions of the unknown for all to see and experience. The map becomes a graphical stand-in (a simulation) for virtuality and is discussed by many of the philosophers in the previous section. Much like Baudrillard's hyperreal and Borges description of the map that becomes so large it becomes reality, "The technology of the map gave to man a new and more comprehending mind .... [which] came to understand reality in the map's terms" (Carr 2010, 41).

At the turn of the twentieth century, electrification allowed virtual worlds to figuratively come alive on the screens of the cinema. The "silver screen" represented a leap of immersive imagination from out of an individual's mind onto a semi-reflective surface. No longer was there a reliance on the spectator to create the immersive effect mentally; instead, the visual field drew them into the electrically rendered representations. One only needs to think of one of the earliest and evocative attempts to immerse the audience in virtuality via film. The Lumière brothers' fabled short depiction of a train pulling into a Paris train station in 1895. The spectators, sitting in their seats in the theater watching this new marvel of technology, became figuratively sucked into the scene by taking the place of the camera via its perspective. When the train approaches the station and toward the audience members, it seemingly bursts into actuality. Virtuality breaking through into actuality seemed possible because the spectators had become so immersed in the fiction that they believed there was no dividing line. This example's effect would later be capitalized on through the implementation of 3D imagery in films of the 1950s and beyond. From shadows on cave walls, to the book, map, movie screen, and finally the digital realm, the immersive qualities of intellectual and representative technologies have been remediated and converged to further pull spectators into the illusory space of imagination. As Murray (2017) reminds us, "A stirring narrative is any medium can be experienced as a virtual reality because our brains are programmed to tune into stories with an intensity that can obliterate the world around us" (124). Each of the mediums above relies on a suspenseful belief on the part of the spectator that the material they are experiencing is real or as real as possible outside of actuality.

The 1980s and 1990s were dominated by tendencies toward fascinations with "virtuality and interactivity" brought about by advances in digital cultures (Munster 2006, 86). For the first time, narratives find a tool for actualization that is interactive and immersive when virtuality becomes synonymous with digital systems. Specific tools of escape, such as computer systems that program and actualize realities within the walls of code, become possible and graft onto the fictions explored in what William Gibson first coined *cyberspace*. In his groundbreaking novel *Neuromancer* (1984), Gibson offers a fully fleshed out exploration of what would become known as VR (virtual reality). Gibson's book marks a definitive shift in science fiction writing as it presents a cyberpunk world where a future humanity becomes mediatized through a merging with technologies. Gibson's virtuality is one where the physical world becomes simulated through digital technologies and is accessible through neural interfaces. He names this immersive digital world the matrix, and the book's protagonist jacks into this simulation by connecting via a brain-computer interface. Like most science fiction, the book capitalizes on an emerging fear and cultural fascination with the rise of computer technology and offers a critique of what the expansion of the digital realm might offer. Gibson's book quickly gained a cult following and influenced a new generation of technologists. It also gave ground to the development of actual VR tech while promoting a counterculture response to the corporatization of contemporary life via hacker culture. Gibson's narrative and conceptualization of cyberspace were quickly followed by the "development of virtual reality technologies, ... [that] extend and literalize the desire to imagine computer networks as narratable, as phenomena that can be experienced subjectively" (Foster 2009, 314). These themes and tools also become the backdrop for the various explorations and expansions of digitalization in representative media discussed below. Gibson's work is prophetic in a fashion as it gave credibility to the actualization of VR technologies that entered the public sphere in the late 80s and early 90s just as the internet was taking off.

Through the last decade of the twentieth century, the exploration of the realms of the virtual and simulation expanded through popular references, computer-generated worlds, and cyber-born identities during the expansion of the internet. The evolution of the internet also saw graphical interfaces flourish via computer screens. The graphical interface gave virtuality an interactive televisual referent that was only possible before through the imaginative rendering of lines of code representing text. As an example, let's return to the previous sequence in the Wachowskis' 1999 film *The Matrix,* where one of the operators stares at a cascading screen full of digital code. The operator sees the virtual depictions of reality in the code by deciphering the semiotic symbolism (reading) of the computer-born language. Before the graphical interface, the imaginative capacity of the human mind was the code breaker who transformed binary digits and textual interfaces into inhabitable and immersive worlds. This mode of virtuality replicated the technology of the printed word, but the visual fidelity of screen-based media (computer, television, film) replaced a necessity for imaginative mental projection. This formulation of being able to become one with the machine or culturally encoded with technology reached a fever pitch in film.

As computers became more common place, the condition of virtuality became a pervasive presence in popular culture. Film narratives such as those in *The Lawnmower Man* (1992), *Strange Days* (1995), *Johnny Mnemonic* (1995), *Virtuosity* (1995), *Ghost in the Shell* (1995), *The Thirteenth Floor* (1999), and *The Matrix* (1999) all played with various themes of virtuality and simulation. In *The Lawnmower Man,* adapted from the 1975 short story by Stephen King, a man with mental and intellectual impairments is selected to undergo treatment in a digital simulator with the hopes of treating his cognitive decline. Once placed inside the virtual world, he quickly becomes much more intelligent and aware to the point that he begins to gain control over the simulation and eventually develops a god complex, lashing out at those running the program. The lines between the virtual and the actual become blurred as he attempts to take revenge on all humanity. The movie becomes a critique of technological progress, and specifically a warning of the potentially catastrophic possibilities of messing with human consciousness using digital technologies and experimenting with merging man with machine. Following a similar theme but inverting the system interface, *Virtuosity* (1995) offers Sid 6.7, a virtual criminal who escapes his simulated confines to wreak havoc on the real world. Again, a warning is given about how society thinks about the blurring of the virtual and the actual through our technologies and specifically the development of advanced artificial intelligence. In these examples, VR technologies are offered as tools of our own human destruction if not managed ethically and responsibly. *The Matrix* (1999), and its subsequent sequels *Matrix Reloaded* (2003) and *Matrix Revolutions* (2003)—and extensive transmedia extensions—offer a world torn apart when machines gain super intelligence and then make humans their subjects. The series capitalized on the "brain in the jar" metaphor to question the very ontology of reality in a world of computer simulation. The original *Matrix* film marks a pinnacle of cultural fascination with simulation and virtual realities. This is not to say that the 2000s moving forward had no engagement with these concepts, but they became less common and often less dystopic. This could be due to an exhaustion with the concepts or due to "end times" not coming into being with the Y2K fever that gripped the cultural zeitgeist at the end of the millennium. More likely, it could simply be that the fears had come true, virtuality had subsumed culture, and society had indeed become conditioned to its effects to a point that technologies were no longer something to be feared. Life was now immersed within the virtual and new formulations of relations would need to be explored. While these films are just a sampling of those grounded in the themes of virtuality and simulation, they are emblematic of a historical time frame where the general populace slowly became increasingly aware that the uneasy liminal space between virtuality and actuality was beginning to overlap.

While the cultural impact of virtuality in the 80s and 90s was profound, there was a lack of broad use of actual VR technologies that these narratives make thematic concerns about. VR tech has its origins in "immersive panoramas and illusionary spaces going as far back as antiquity" where the use of scale and perspective gave the viewer the feeling of being engulfed by the artwork (Dixon 2007, 363). In 1838, Charles Wheatstone developed the first tool that could transform 2D artwork into 3D through the use of stereoscopic imagery made possible by special viewing devices. The device would segregate each eye and focus it on an individual image. When the spectator saw the combined images, their mind turned the two into one three-dimensional object. This use of stereographic imagery is the basis for 3D movies and the VR head mounted display (HMD) technologies we have today. These early examples of virtual realities come close to fitting the first axis of historian Howard Rheingold's three-part taxonomy of VR representation:

1 "Immersion, being surrounded by a three-dimensional world."
2 "The ability to walk around in that world, choose your own point of view."
3 "Manipulation, being able to reach in and manipulate it" [the virtual field].
(Rheingold 1991 quoted in Dixon 2007, 364)

Each example allows a form of mental processing based on making the person's mind feel immersed by blocking out the actual reality beyond the image source. One of the defining traits of VR is tricking the "brain into believing something is real, even when it isn't" (Virtual Reality Society).

David Chalmers (2022) suggests that VR offers an "immersive virtual world" which becomes "fully immersive when one is immersed in the virtual world *with all of one's senses*, experiencing it just as we experience the physical world" (39, my italics). Through most VR, the spectator is offered a tool to enter this virtual world, gaining the sensory possibility of accessing virtuality mainly through ocular and aural sensing. Starting with the stereoscope device, virtual realities began to take on much more advanced technological formats, where the tools create a simulated environment for the user (Giannachi 2004, 133). Before advanced computer processing, the second two parts of Rheingold's taxonomy required large-scale pictorial projection and physical placement within a designed environment to achieve their results. This is similar to the CAVE style of VR environments tested in the early 90s where physical space is augmented with projected televisual technology versus simulating space with digital technologies. These aspects of virtual immersion are common in the simulator technologies that first were developed by the military to train pilots, which then quickly moved over to the commercial realm and eventually morphed into entertainment formats. For example, in the late 80s, nearly every suburban mall in America had a simulator ride. These simulators would replicate the experience one might find at a theme park or as a sports daredevil. They often looked like small escape shuttles from sci-fi television programming and could simulate the experience of hang gliding or a roller coaster by matching a first-person film perspective with mechanical movement created within a closed setting. The combination of the physical bumps and dives with the close proximity to the screen gave the impression of the action being filmed. These types of simulation or VR experiences were simply films shown in an enclosure manipulated with electromagnetic technology. They had yet to deliver the ability to step into and navigate virtual space per Rheingold's two second criteria. It would take more technological advances to do so.

In Kevin Kelly's (2016) book *The Inevitable*, he details how the potential of VR technology one might see today has long been in the works. For him, "virtual Reality is a fake world that feels absolutely authentic" (211) and represents the pinnacle of artistic representation. VR is the ultimate artistic creation, as it renders the world of dreams using digital tools and processes, giving its user an approximation of reality amped up to the nth degree, offering a world where the spectator can escape the limitations of the real world. As an early and active member of the cyber community and the founder of *Wired*, Kelly was introduced to some of the first consumer friendly versions of the technology in 1989. What we have come to understand as VR today did not truly begin to take shape and inform the general populace's understanding until the mid-80s when Jaron Lanier and Thomas Zimmerman formed VPL research to develop commercially available products. Kelly was introduced to these new affective technologies through Lanier's research, and as an early cyber enthusiast, he helped promote and support the cultural obsession with digital realities. While the foundational technologies had been around for decades, they were largely kept away from the public eye.

Since the late 1960s, HMD technologies that were connected to computer processing systems instead of a camera or film image were used for research and training by universities and the military, but they were bulky and incredibly costly. To give an example of the scale, one of the earliest was nicknamed the Sword of Damocles due to the size and way the user interfaced with the tech. The user would don the helmet in the center of the room where external head tracking sensors could monitor the user. The helmet was so heavy that it had to be held up and controlled by a mechanical arm which could also assist in the movement. This size of the arm gave the impression that the user was in danger of having their head removed, hence the nickname. The size and weight of HMDs have been one of the most complex problems VR technologists have had to manage, and not until the last few years has computing technology become small and light enough to allow unimpeded user movement. Another challenge was visual fidelity. Early HMDs could not generate optically clear simulations that could truly immerse the user in the virtual representations. To address this problem, VR researchers believed that pairing other sensory inputs might alleviate the fidelity problem. Lanier sought to pair the HMD technology with a haptic device which would give the user the impression of control and agency over virtual realities. He developed the first consumer *data glove,* a tool that allowed the user to touch and control objects in digital space. Having experienced Lanier's products, Kelly explained that the "goal of VR is not to suspend belief but to ratchet up belief" (212), to make the user feel so immersed in the experience that virtuality becomes more authentic than actuality. This matches Giannachi's (2004) understanding of spectator experience within VR: "Virtual reality is in a paradoxical relationship with the real. On the one hand, it is part of the real; yet, on the other, it must be constructed as different from the real to be perceived as separate from it" (123). The goal for Lanier was to use haptics to link the actual to the virtual already created using the optical and auditory components of the HMD. They understood that virtuality becomes most real for the spectator when coupled with affect. Virtual immersion via only visual or auditory stimuli will eventually wear out without another form of affective input, and this means activating the physical world via haptics. The mediatized spectator's body is a sensory input machine working in multiple operations and combinations between the visual, auditory, olfactory, haptic, and imaginary world of virtuality. It is the combination of these elements that allows one to comprehend the experiential nature of virtuality.

Lanier would become famous three years later for coining the term virtual reality that would commonly be known as VR. Lanier's data glove represented a cultural shift in how virtuality was seen by the common person. Haptic gloves modeled after Lanier's became a cultural phenomenon and fantasy of the era. I remember desperately wishing to own Nintendo's Power Glove after watching the movie *The Wizard* (1989). The Power Glove was a motion-activated controller for the NES system. The glove was unveiled as next-gen tech that materially connected the user to the virtual domain of the game. In the movie, the glove appeared in a dramatic reveal where the story's antagonist, Lucas, produces a silver box with the controller inside. The "wizard" and his cohorts are astounded and intimidated by the technology as Lucas dons the glove and masterfully completes a level of *Rad Racer* using a motion as if he is holding an invisible steering wheel. The scene ends with Lucas stating, "I love the Power Glove. It's so Bad" (Holland 1989). Truthfully, the controller was bad, but not in an ironic sense. While the controller ended up being a commercial success, it had very limited functionality with most NES games. Critics bemoaned the movie as being a long promotion video for Nintendo (Greene 2015), but it helped usher in a cultural realization of the possible overlap between virtuality and actuality using technological tools. With the glove, an idea of being able to embody action in virtual domains was imprinted on the cultural zeitgeist of the time. Both Nintendo and Sega attempted to develop a gaming system utilizing an HMD system paired with a controller, but neither were successful enough in testing to release broadly. While both companies were already setting the foundations for gaming culture with their home entertainment systems, they did not have technology enough to bridge the challenges of immersion through affective sensing. Instead, they would capitalize on the experience of game play to capture the attention of users. The promise of escaping reality via technological immersion would have to wait until technology could generate the field of possibilities explored in the films discussed previously.

In 1989, Kelly (2016) remarked how he believed that, "VR tech would be ubiquitous in five years or so" (215), and more than a decade after his introduction to the tech, Giannachi (2004) extolled that VR was "one of Western society's primary tools to present and advertise itself, reflect on itself, create experience of itself that furthers knowledge and thus becomes art" (124). Both perspectives were utopic and captured the feeling at the end of the century where virtuality became embedded in society through digital technologies. Even so, their predictions and assertions would take decades to become possible for the average consumer. The truth is that the technology for enough visual fidelity to truly immerse the user would not become readily available until the mid-2010s. By this time, all reality had become subsumed by virtuality, and escape into virtual realms was less of a dream but a common daily occurrence, just not in the way Lanier, Kelly, and Giannachi might have hoped.

While the tech community was waiting for advances in screen technologies that could live up to VR's potential, a wave of ubiquitous computing would instead create digital networks to engulf actuality in virtual technologies as opposed to putting humans inside of virtual constructions. This meant that many forms of contemporary would have to rely on haptics for immersion by affective presence. The goal of VR is to bring the spectator as close to a new reality as possible. This reliance on presence is why the immersive capacity of current VR needs to couple to some form of emotional or haptic hook. Next-generation VR makes digital simulations more real and encompassing of the perceptual apparatus by capitalizing on the synchronous duality of both body and mind perception. VR video games also capitalize on presence by inducing emotional response paired with interactive tactics of gamification and haptic feedback via controllers to induce the feeling of actuality in the virtual space. This is why the most memorable VR experiences include either horror or physical action. With full-body suits packed with haptic sensors or personalized construction sites for physical referents, VR gains the potential to bridge the gap between virtuality and actuality. But until these suits are integrated into society, people will continue to navigate the worlds of possibility made real through our bodily interactions. Without interfacing with the body, VR has less capacity to become actual. That is until the entirety of the actual world becomes wired and connected, making the actual a virtual world. Causey (2016) sums this up when he states,

> It is, of course, the human-computer interface (HCI) that dictates the production and aesthetics of virtual spaces. As technological advancements move us past the constrictions of keyboards, head-mounted displays, and data-gloves toward more open, immersive spaces of ubiquitous computing, the interface will recede in a co-mingling of body and machine. (50)

When a spectator gains an ability to perceive virtuality in some other form than the primarily visual, they can enter a non-distinguishable two-fold immersion between both virtuality and actuality via embodied affect. This capacity to traverse both Reals opens up multiple questions about how one understands their own sense of being in the work due to the fact that virtual realities are also "about the definition of the self and the relationship of the body to the world" (Bolter and Grusin 1999, 166). The mediatized spectator must navigate these Reals under the condition of virtuality.

To help pilot and ground artists and researchers' understanding of spectator experience, they must then focus on the body and its perceptual faculties. In the time when technology was trying to catch up to the techno-fueled dreams of the 1980s, theatre and performance artists began to increasingly capitalize on the desire to transcend physical realities using immersive aesthetics.[2] Replacing the space of the imagination and the screen, the feeling body would become the site of virtuality through physical immersion in scenic space. Immersive embodiment has become the way spectators are increasingly accessing virtual realities today. Contemporary technologies constantly require us to imagine ourselves being transported into digital and televisual spaces through our corporeal actions. Visually, we embrace virtuality via the many screens that encompass and graft onto our bodies; aurally, we access virtuality through the constant streams of digitally delivered music; and tactilely, we embrace virtuality every time we swipe and tap on our smartphones and tablets. The body portion of our perceptual apparatus becomes imbued and intertwined with affective potential, making our sense of being one that is immersed in virtuality.

Immersion: A Structure for Personal Affect and Sensory Experience

Let's return to where we started this chapter by asking you to imagine the world around you. Do you now have a better sense of what is real? Or are you now doubting the reality you live within? In the previous section, we ended by focusing on embodiment as the grounding object through which we might be able to describe experience within virtual realities. Immersion is the frame through which one might define their actuality within the world. Running counter to Descartes, the body is the technology through which the virtual—the space of possibility and potential—becomes actuality, or what some philosophers might call the real Real. If the goal of an artist is to immerse one in virtual space/time, it is through bodily response and affective perceiving that we can best understand the spectator experience. The architecture of immersion is therefore a framework through which one determines what is real and what is not via their individual subjective position. This remaining section will focus on the concept of immersion in media and how it has been used by scholars and artists to define a specific type of experience. I focus specifically on how agency manifests for the spectator and what form of exchange is offered through acts of spectatorship.

According to Josephine Machon (2016a), the term immersive is most often used to "define a style of performance practice and applied to diverse events that seek to exploit all that is experiential in performance" (35). She also adds that the term is "now used freely (sometimes excessively) to describe contemporary performance practice involving a visceral and participatory audience experience with an all-encompassing, sensual style of production aesthetic" (25). Through her books *Immersive Theatres* (2013) and *(Syn) aesthetics* (2009), Machon ushered in a new era of critical and popular scrutiny (and arguably its increasing use and popularity as a practice) about immersion in theatrical performance. Immersive is often casually used as an adjective to describe many interactive, site-specific, and play-based forms of performance. The term has become a marketing catchphrase that is deeply enmeshed in the neoliberal experience economy (Alston 2016). Because "immersive theatre" is used for a variety of practices that employ modes of experience, I find it a useful place to begin discussing mediatized spectatorship. As I will later discuss in brief, there are some political and economic determinates that have contributed to the increasing pervasiveness of immersivity, but the primary purpose of this section is to discuss the underlying operation of immersion as a form of experience that relies on affective response and implied spectatorial agency. By doing so, I can advance through the rest of this project describing the other architectures of exchange that are often combined with immersivity to create experiential spectatorship. For the purposes of differentiating immersion from the other architectures in this book I will define it in this manner: Immersion is a mode of exchange that allows a spectator to be thrown into a fictive world, giving them the impression of being a member of that world and allowing for heightened levels of perceived agency based on the beingness with(in) that world.

Multiple ways of defining the immersive experience have emerged over the years and with variations based on the objects that it occurs through. Janet Murray (2017) describes immersion using the metaphor of water:

> Immersion is a metaphorical term derived from the physical experience of being submerged in water. We seek the same feeling from a psychologically immersive experience that we do from a plunge in the ocean or swimming pool: the sensation of being surrounded by a completely other reality, as different as water is from air, that takes over all of our attention, our whole perceptual apparatus. (124)

Nearly 20 years later, Machon (2013) also uses the water metaphor to invoke the immersive capacity of experiential theatre.

> Remember what it is like to be immersed in water; to lie back slowly and put your head underwater in the bath. The muted sensation of being submerged in another medium, where the rules change because if you were to breathe as normal your lungs would fill with water [...] We understand what it is to be immersed in water; the action of plunging your whole body into an alternative medium and its subsequent sensations [...] I know when I have experienced a wholly immersive event I am totally submerged in it for the length of time that the event lasts, aware of nothing other than the event itself and only actions, feelings (both emotion and sensation), and thoughts related to event are of consequence in that time. (xiv–xvi)

In both examples, the immersed spectator or *immersant* must cross a threshold where they lose some sense of their own agency to act, to remain the person that exists outside the immersive environment. Inside this space emerges a new sense of agency that is unique to the individual operations and aesthetics of the immersive environment. In an environment where one is submerged/immersed, one must learn to "do the things that the new environment makes possible" (Murray 2017 124). Being simply immersed is not enough, there must be a perception of being able to do something beyond being submerged. I argue that a sense of agency that is perceived but actually non-existent beyond individual affect is a crucial element to the experience of immersion. Remember that agency is defined as the ability to do something, to act. In the case of immersion spectators simply respond through embodied sensation. Mediatized spectators seek to engage with the virtual space of performance by interacting and being interacted with, but not all immersive environments are interactive. Just like the pool in which one is submerged, learning to swim is not inherent; one can sink, swim, or float.

Murray's (2017) discussion of immersion focused on form and conventions within cyber-assisted computer narratives where "digital media take us to a place where we can act out our fantasies" (124). Murray was at the forefront of scholars defining the societal impacts of digitalization and specifically how digital products were changing the landscape of literature at the turn of the twenty-first century. Murray's focus on immersion is influenced by the multiple theorists mentioned previously and considers the rise of digital narratives as they connect to digital gaming and VR technologies where the terminology first gained traction. Usage of the term immersive in print stayed relatively flat through the 1980s, increasing only twofold, but in 1992—the year *Lawnmower Man* premiered—it began to pick up traction and exponentially increased through the decade by 956 percent. Between 2000 and 2019, the term increased another 645 percent with a steep increase between 2015–2019 (see Figure 2.1). The early increase of frequencies coincides with the scholarship on virtuality, digital space, and simulation technologies. The most recent increase coincides with a second wave of interest in the subject by theatre and performance studies scholars, alongside the resurgence of interest in mixed-reality products. Murray and Machon represent two central pillars in the shifting interest in the vocabulary of immersivity.

If one were to run a Google search for the adjective immersive, they would most likely see this definition, "(of a computer display or system) generating a three-dimensional image which appears around the user" (Oxford Dictionary of Language). This usage connects to its common usage origin related to technologies of simulation shown in the tracking data. A further search connects the term to artistic domains or media, film, and theatre and is defined as "seeming to surround the audience, player, etc. so that they feel completely involved in something" (Cambridge Dictionary). And the Merriam Webster's Dictionary defines it as: "providing, involving, or characterized by deep absorption or immersion in something" (Merriam Webster). In each, there is an emphasis on the creation of an environment—either virtual or actual—that envelopes a person and gives them the sense of belonging with(in) that space. These definitions also point to how the terminology shifts between the cultural spaces of technology and artistic production. At a time when there seemed to be exhaustion with the

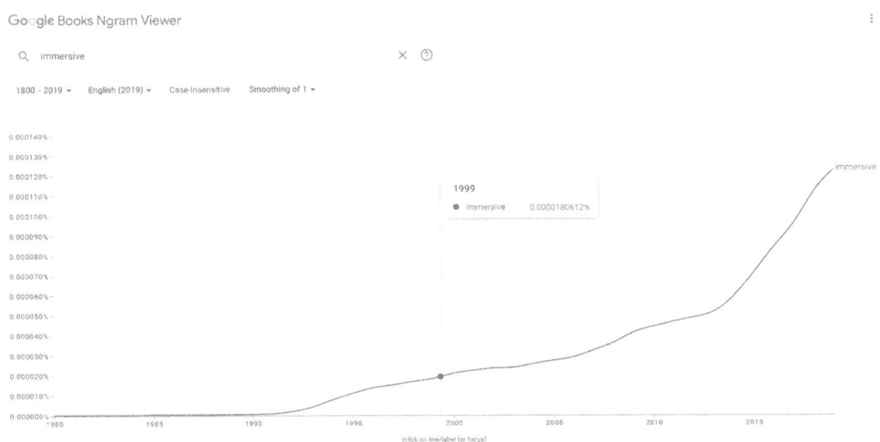

Figure 2.1 Increase in the term *Immersive* from 1980 to 2019. Google Ngram Viewer.

virtual and its immersive capacities, the usage shifts into a new cultural space, giving rise to the trend toward immersive theatre and performance practices. These practices then begin to become the primary objects of analysis in the early 2010s and begin to divorce themselves from the technological variety. This points to how virtuality was embraced and explored through technological products but then shifted to paradigms of non-technologically situated embodiment within performance venues. This is not to say that scholarship that combines the two was formally separated, but it marks a divergence in focus.

Within theatre and performance studies scholarship, there exists a focus on the technologically mediated strand of immersivity (Birringer 2006; Dixon 2007; Giannachi 2004; Kattenbelt and Chappel 2006; Bay-Cheng et al. 2010; Benford and Giannachi 2011; Klich and Scheer 2012) informed by theories from technology and media studies scholars (Laurel 1991; Murray 1997; Bolter and Grusin 1999; Hayles 1999; Ryan 2001; and Massumi 2002). This body of work sits alongside a focus on embodiment and agency within immersive physical spaces and situations (Machon 2009; Machon 2013; White 2012; Lavender 2016; Alston 2016; Frieze 2016; and Westling 2020) informed by cultural and artistic perspectives. The primary shift is from a focus on the technological products and cultural conditions that produce the immersive affect as a form of interactivity to the situation and experience of the spectator that arises through the creation of a physical immersive environment. These two strands have often been so wide and diverse that they allowed the encapsulation of all experiential forms to become named immersive. Few scholars have focused intensively through full-length book form to bridge the immersive capacities of technological tools/environments (virtual) and physical environments/situations (actual) through embodiment. Jennifer Parker-Starbuck's (2011) *Cyborg Theatre* and Liam Jarvis' (2019) *Immersive Embodiment* carefully weave both strands together and serve as bookends to most work done in the last decade on the subject. Rose Biggin's (2017) *Immersive Theatre and Audience Experience* stands out from the pack as it deeply mines the vast continuum of perspectives on immersion from media, technology, theatre, audience, and game studies and is crucial reading should one want to best define how immersion is and isn't a form of interactivity. The drawback to her work is how it primarily focuses on the work of Punchdrunk for her case studies which negates in some capacity the excellent work of surveying the various fields of study.

In an effort to refine the usage of the terms Machon (2016a) seems to shift some of her own terminology concerning spectators in immersive productions when referring to spectators of immersive events as *interactors*. She explains:

Immersive work physically takes us into the world of the play, rather than inviting us to spectate and comprehend it from a distance in an auditorium. In most immersive practice, the space is integral to the experience. The audience is not separated from it but in it, of it. Interactors are surrounded by it, dwelling in it, travelling through it, which ensures some sense of 'rootedness' in the world of the event is actively felt. (44)

This shifts the focus back to the body of the immersant through her early work on synaesthetics to show how immersive work is one that "require[s] a sensual comprehension of the work in the immediate moment that influences any subsequent analysis of each event" (Machon 2016b, 31). One should not conflate an analysis of the aesthetics of an immersive event with the individualized experience of a spectator interacting with(in) that event. When one considers the spectator, they must also then consider if their analysis is one of first-person perspective as the immersant or from the third-person perspective as cataloger of other's experience.

Murray (2017) also uses the term interactors when discussing those who engage with virtuality via digital immersion. Their use of interaction seems to run counter to some of their notions of immersion as being simply submerged in a world. Interaction comes from the ability to sense the world they are submerged in but not necessarily act upon those senses. The immersant as interactor seems to point to an attempt to reconcile the two strands through the body of the spectator. Biggin (2017) offers a detailed argument for how immersion is not necessarily a form of interactivity but rather an aesthetic feeling that arises out of a variety of different forms of interactivity. She cites Dixon's (2007) four-part schema for interactivity in digital performance—Navigation, Participation, Conversation, Collaboration (564)—alongside five characteristics of multimedia—integration, interactivity, hypermedia, immersion, and narrativity (Packer and Jordan 2001, xxx) as models for thinking about the ways the digital products work with the spectator. These are offered as a way of defining immersion as separate but linked to interactivity which is the prevailing methodology for those covering the technological spectrum of research on immersion. These lines of thought map onto the conversation about how digital and virtual products create an immersive affect with and on their users. This is a novel nuanced approach where immersion is a form of exchange between the virtual construct/environment and the spectator's perceptual apparatus. The interaction is one prior to conscious action and is based on sensory input received which becomes interactive within the feeling body. This is different from the form of "cognitive immersion" based on "brain activity" and instead focuses the spectator's attention and consciousness toward the physical "here and now of the performance" (Biggin 2017, 61).

Interaction in most theatrical immersive experiences begins with bodily perception through various levels of affect. This coincides with Alston (2016) when he states, "Affect is used as a channel that connects audiences to a fictive world, making them feel included as a part of a world that they help to produce. The aesthetic experience that results from their productive endeavor is what tends to take aesthetic precedence in these performances" (175). The spectator has agency to create meaning by harnessing the affective possibilities of its perceptual apparatus. Agency is a crucial element for the creation of meaning in this aesthetic and becomes a central concern when discussing all the architectures of exchange. Before the turn of the new millennium, Baz

Kershaw (1999) discussed agency as a dynamic component of *"an aesthetic of total immersion"* (194, italics in original) in participatory performance practices. I find it useful to mark this contribution to the theoretical landscape of performance because it foregrounds the necessity of action beyond submersion or encompassment in an immersive event.

Immersion in actuality is made possible via affective registers, and immersion in virtuality is made possible by the impression of these registers developing in the perceptual apparatus of the spectator. There is a doubleness at work in the relationship between virtuality and actuality. When virtuality subsumes actuality, the latter becomes the limits of the former. Massumi (2002) states that "Affect is *the virtual point of view* ... Affects are *virtual synesthetic perspectives* anchored in (functionally limited by) the actually existing, particular things that embody them" (35, italics in original). A body that is immersed in virtuality (in this case, a technologically augmented VR environment) grounds the experience of realness via affective response. While technologically advanced societies may now already be immersed in a condition of virtuality, I argue that a mediatized spectator remains an embodied entity who must navigate the difference between the virtual and the actual. This runs counter to the argument of many theorists of media and virtuality. Anna Munster (2006) explains that early notions of the virtual in digital culture "promised to leave the body and its 'meat' behind, as minds, data and wires join together in an ecstatic fusion across the infinite matrix of cyberspace" (86). Instead, virtually is embedded with(in) the mediatized body, while this body is also embedded in networks of virtuality. An embodied mediatized spectator is more in line with Hayles (1999) formulation of the posthuman whose body has entered the informational flow but has yet to be de-materialized into pure information. A mediatized spectator is then a perceiver of the in-between, navigating the cognitive realities of virtual information flows and the affective responses of the physical body to its environment.

2.1   The Feeling Spectator and the Affect Economy of Immersion

In this hinge section, immersion is shown to operate in ways that pull spectators deep within their own imagination to virtually augment reality "on stage." The Instances explore different experiences of immersion and their connection to the various representations of virtuality. Agency in immersion often surfaces primarily through bodily affect, though it also manifests in other ways when the immersive event uses other forms of exchange. The primary mode of exchange highlighted in the *Architecture of Immersion* is sensual affectivity. Immersion in both actual and virtual domains relies on a state of sensory engulfment by the spectator's perceptual apparatus that allows the spectator to create forms of meaning where they feel as though they are a crucial part of the event with agency, even when this "is not true." This mirrors the relationship between virtuality and actuality. The feeling of agency is usually only expressible through individual meaning-making versus direct action, unless one of the other architectures of exchange is included. When paired with immersion the other architectures allow the experience to become more than a passive dip in the pool.

While I am often critical of the casual use of the term immersive and find it necessary to further develop a methodology for how to analyze and apply it, it is an integral part of the mediatized condition as brought about by virtuality. Before moving on to the Instances, let's highlight part of Machon's (2013) argument about immersion: "I am now certain that 'immersive theatre' is impossible to define as a genre, with fixed and determinate codes and conventions, because *it is not one*" (xvi, italics in original). There is no such taxonomy as immersive theatre, but one can define an immersive experience or as Biggin (2017) argues, an immersive aesthetic. There are multiple immersive theatres and likewise multiple immersive media. Each is defined by their affective experience and connection to other architectures, and therefore, it is beneficial to approach the term immersive as an experiential descriptor based on an individual form of exchange.

Immersion is therefore a specific architecture of exchange that often relies on a combination of imagination and affective (virtual and material) inputs. Affect emerges through the many different sensors contained in and on the human body. A connection to the digital realm and an affinity toward a condition of virtuality have helped tease out the dynamics necessary for a mode of affective exchange to emerge through that body. Immersion serves as a helpful and necessary foundation for mediatized experience as it is connected to the technic of virtuality which has an overarching relationship with digitalization. Utilizing immersivity as a frame for spectatorial experiences highlights how embodiment is necessary when considering overlaps and divides between virtuality and actuality. The terminology is also useful when anchoring the body portion of a person's perceptual apparatus to both the divide and the overlap. The technology to transcend material existence does not yet exist and therefore mediatized spectators will continue to rely on their bodies to work as conduits for immersive experience.

Instance: Enter the Void—the Hyperreal Dimensions
of the Haptic/Virtual Interface

I've just strapped on a twenty-pound contraption that confines my torso in a technological cocoon of wires, mesh fabric, and plastic. On my chest plate are various nodes that I later find out are haptic transducers that pulse and vibrate based on the actions I make. On my back is a shell that contains my own personal, world-generating computer. I have also donned a helmet with a built-in electronic visor that digitally creates the visual field I am about to enter virtually. After strapping myself into this body rig, the proctor hands me a plastic molded "weapon" I will use to bust some ghosts. In the actual world, the weapon has two buttons and a slide mechanism like that on a shotgun; in the virtual world, I am holding a realistic-looking blaster extension of a proton pack. I'm ready to experience *Ghostbusters: Dimension*, the immersive virtual reality attraction at Madame Tussaud's in New York City.[3]

Ghostbusters: Dimension was one of the first commercially available immersive VR experiences from the Utah-based tech company The Void. The attraction launched on July 1, 2016, at the wax museum as part of a larger interactive themed attraction connected to the launch of the 2016 remake of the iconic 1980s film. The larger experiential museum-like attraction has one interact with technological devices such as the famous ghost containment unit and encounter holographically generated ghosts in a recreation of New York City subway tunnels and buildings. As a special add-on to the simulated movie experience, I purchased the ticket to the VR experience. At this point in my research, I had only personally experienced VR in the mode of Google Cardboard or similar smartphone-based stereographic headsets. The press release for The Void's attraction referred to its version of simulated reality as "hyper reality" that operates on the same functionality of "virtual reality" systems but is made even more real "by combining physical sets, real-time interactive effects, and virtual reality. This allows participants to not just watch a movie or play a game, but to live them" (The Void 2016). The attraction did just what its creators promoted, with caveats. After fully strapping on and jacking into the experience, I am instructed to begin once I see the light above me turn on.

Inside the virtual world projected onto the surface of my goggles, a door is directly in front of me with a canister-shaped light above. The light turns bright white, and I instinctively reach forward toward the door. I am shocked and drawn further into the simulated reality when my hand meets an actual doorknob that I must turn and then push to open. The system has upped the ante on the experience of virtuality by linking actuality and virtuality via affective sensuality. As I push through the door, I digitally enter a non-descript apartment room with a small sink at the far corner. I reach out to feel the real walls projected in my visor, the tap handle of a sink, and other physical aspects that ground me in this reality. I *feel* like I am actually in this digital space. Shortly after doing so, cute and seemingly non-threatening apparitions visually appear out of nowhere. I instinctively address my proton blaster in their direction and pull the trigger. Nothing happens! The blaster does not shoot, and I begin to wildly flail about the visual space, trying to zap the specters with a terrifying jolt of proton electricity. Even though I am receiving haptic feedback via the chest sensors when the ghosts fly at me, I cannot fully connect because every time I pull the trigger, there is no response. My technology's defect has relegated me to that of a virtual tourist left to merely float along in the river of digital imagery.

Soon after, another larger and grumpier looking ghost soon appears and taunts me before breaking through a nearby wall. I follow in pursuit and am led to an elevator where, nearby, another ghost sings a creepy lullaby. I'm told via my headset that this apparition was an inhabitant of this building years ago but went missing. Shortly after entering the elevator, the doors slam shut, and the ghost comes screaming at me and then through me. As she transports through my digital body, I feel a cool and moist spray attack my actual skin. Is this what it feels like to be briefly possessed? The experience induces a mild fright and the sensation of the spray, while disconcerting, makes me feel more connected to the virtual world. The carriage passes up the elevator shaft that I see through the open top. The elevator is like one of those steel cage varieties seen in the lobby of some 1920s gangster film. Visually, the ride up the elevator seems real, but there is no feedback like one would feel in a real elevator, specifically one of this age and technological capability. Without any haptic sensory feedback, I am again reminded that this experience exists primarily in the relationship between my eyes and my mind.

At the top of the elevator, I enter a hallway where a large part of the structure has been torn away to expose the exterior of the New York City skyline. The sky is "on fire" with an interdimensional gateway opening. I step out onto what looks like a painter's walkway or window washer's scaffold, and as I look down, I can see the city streets forty stories below, inducing a slight sense of vertigo. As I move, the walkway stutters and shakes, making me reach out for the virtual handrails. Luckily this "room" has an actual set of rails for me to grab onto or I may have fallen flat on my visor. Immediately, a throng of gargoyles come to life and start rushing toward me. I try again to zap them with my proton blaster; again, no result. The experience again begins to simply feel like I am watching a 3D movie where images can come rushing at me in my visual field, but because I cannot truly interact with them, the experience feels wanting. The gargoyles fly away, and I enter another room where the ghost lady from before accosts me and taunts me into a fight by throwing chairs, bricks, and other items at me. My haptic vest is going crazy, and I feel as though a physical being is trying to harm me. The blaster still doesn't work, though. As my exasperation reaches a fever pitch with not being able to do anything but get pelted by digital objects, the good old Stay Puft Marshmallow Man appears and angrily begins to swipe at me through the gaping hole in the room's wall. I'm instructed to burn him up with my blaster, which I of course cannot do. After a minute or two of awkwardly dancing with this twenty-story confectionary wonder, he disappears with a whiff of marshmallow scented abandon. Weirdly, I actually smelled its departure.

From there, I exited the room and was led by the proctor back into the staging area. It was an interesting experience, but the lack of physical interaction during the most intense moments of immersion in the narrative left me very dissatisfied. I didn't feel like the encounter was any more real than a computer simulation. I explained this to the proctor and, after checking some settings, I was given a second chance to experience the event. This time through, I got what I was promised. With the proton blaster now working, every tactile and haptic response brought me closer to the ghosts and allowed me to move in sync with them. I was given the ability to experience them as if they were actually there with me in the room. Capturing the ghosts was met with a trembling and pulling sensation of the proton pack. The feeling was like that of having a fish caught on the line fighting for escape. The sensual and affective qualities of the equipment layered with the proximity and tangibility of the physical space convinced my perceptual apparatus into believing that the virtual world was the actual world. By pairing affective sensing with mental processing, I could suspend any disbelief in the fact that the ghosts were nothing other than digital simulations. I forgot that I was wearing a vest and goggles. I forgot that the ghosts were digital cartoons. I forgot that the suspended walkway was just a contraption made to fool my sense of balance. And when it came time to fight Stay Puft, my proton blaster imparted a satisfying toasting burn to his jolly face. That smell of marshmallow wasn't the trace of his escape; it was my purposeful immolation of his facial features into a layer of campfire S'mores. This time after exiting The Void, I was a sweaty mess, my heart racing, and my adrenaline pumping as if I had just actually fought off the attack of an angry enemy from the netherworld.

Again, I am strapping on a technological contraption meant to make virtuality into actuality. This rig, however, does not contain the same level of sophistication as seen in the previous example. My VR rig is a simply constructed plastic visor with a cord running down my back, with a control switch at my side where I plug in my headphones. In my hand is a rectangular plastic contraption that resembles a folded piece of PVC pipe with a toggle-controlled button at the thumb, a trigger for my forefinger, and at the end of the fold another toggle button intended for my non-dominant hand. At the end of this simply constructed virtual gun is a soft rubber ball that glows with a blue or red light. The contraption is PlayStation's Aim Controller, packaged with the 2017 VR space shooter game *Farpoint*. Like the experience of *Ghostbusters: Dimension*, this consumer-grade VR experience launches me into a visually expansive world intended to immerse me in the sights and sounds of another reality. Unlike *Ghostbusters*, there are few haptic augmentations in this world. Without the full range of haptics and tactile response, I must rely on in-game responses for engagement and connection. Gamification becomes the additional operation for this mode of immersion (Figure 2.2).

As an immersant in this virtual space, my task is to wander about a forsaken desert planet to search for my lost crew of intergalactic travelers. The realism of the world is incredibly well constructed visually considering the technological capacities of the gaming system extension. As I walk across the landscape, I can look over my shoulder or to my side and see my own shadow traveling with me. This visual fidelity is at first remarkably effective at tricking my mind into believing I am actually on this forbidden planet. The physical features of the planet have enough detail to give me the impression

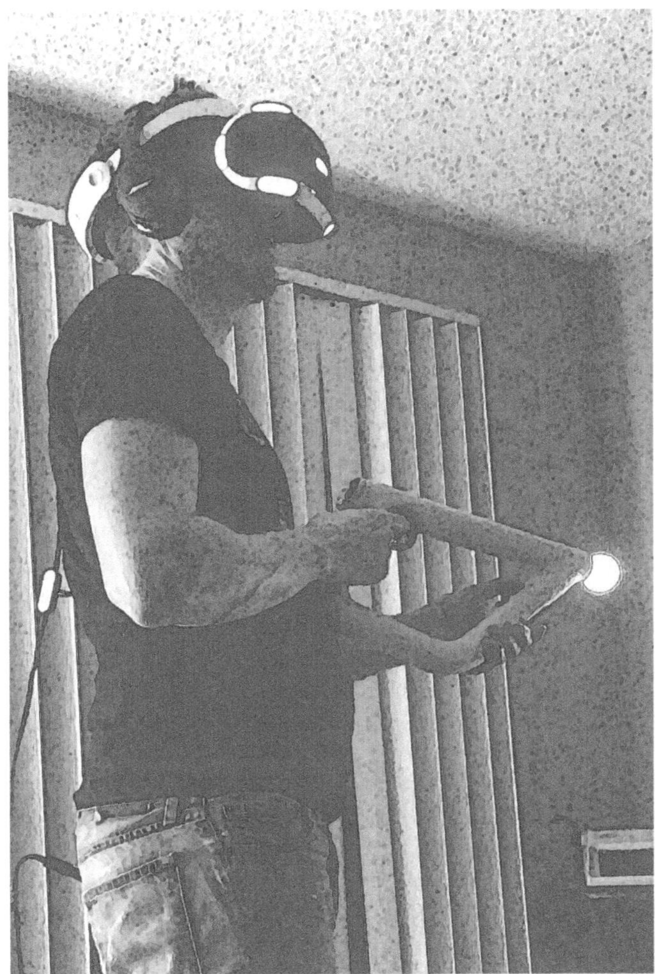

Figure 2.2 Aim Controller and PS4 VR system in action.

that my body has been transported into the digital realm, but it is a mere mental impression. There is an obvious disconnect between my body and the space. To walk forward, I use the thumb toggle on the controller, and it is as if I am simply floating along, even though I can look down and see my virtual feet moving. I find myself at times walking in place trying and sync my actual body with my virtual body, but it has a limited effect. The problem with the disconnect between these two bodies and realities becomes even more apparent when I begin to become mildly nauseous while moving about the planet. This primarily happens when I try to turn virtually. This phenomenon is often referred to as "virtual reality sickness" and is like the affective situation of car or sea sickness. It occurs when the visual field of reference one experiences doesn't match up with the corporeal frame. In the game, if I try to scan around the space using the thumb stick instead of turning my head, it is as though my mind cannot understand how I just whipped around in one direction without actually moving. To combat the nausea, I find it helpful to create a visual frame of reference or static point. This is the same as the directions given to dancers who spot a static place ahead of them during their turns. It helps alleviate dizziness, which is a similar corporeal response to VR sickness. In the game-world, I can hold the gun up to my shoulder like a real rifle, which then brings the gun's sights into my visual field and allows for a static visual frame of reference, though it exists digitally.

The gun/controller is the primary tactile and haptic connection allowing me to ground the immersive affects in my body. When making the game, the designers at Impulse Gear decided that the way the controller is used would set it apart from other PlayStation VR experiences (Tach 2016). The way one uses the sights to create a visual frame of reference is a primary consideration along with how one changes out weapons during gameplay. In most non-VR shooter games, weapon exchange comes by way of a button press. In *Farpoint,* you must bring the controller up and above your shoulder in a motion as if you are grabbing a different weapon from off a backpack style holster. The motion gives the impression of actually having a laser rifle, and not just a simulated weapon. The inclusion of the controller gives the player the affective impact necessary to help ground reality, and, as I explained above, using it tricks the mind into believing that what is created via visual and auditory stimulation is an actuality.

Other than the Aim Controller's connection to the virtual realm, few other affective connections to immersion exist. Instead, emotional response replaces the need for material elements like the purpose-built rooms of *Ghostbusters: Dimension.* This aspect of the game is played upon to great effect. While moving across the arid digital landscape, one encounters a bevy of mutant spider creatures. There are four primary types: (1) drones that pop out of the sand or crawl down from the rock formations; (2) soldiers who throw acid bladders at you; (3) giant rock spiders that charge at you; and (4) what I call diggers, who, upon seeing you, burrow underground to then later pop up out of the sand at your feet to spit acid in your face. The diggers and the drones are by far the most terrifying, as they play on one of the primary emotional registers of all 3D reality imagery: Visual proximity. Like the filmed train virtually hurtling out of the screen mentioned earlier, these creatures love to attack by propelling their arachnid bodies at your face with extreme speed. At first, these visual frights are nearly impossible to play off as only virtually constructed digital images.

Bolter and Grusin (1999) ask, "If virtual reality can evoke emotions, how can our culture deny that the experience of virtual reality is authentic" (165)? Authenticity of embodied experience develops in *Farpoint* primarily by ratcheting up the belief of reality via emotional response. In this environment, where there is very little haptic connection between the body and the virtual space, immersion develops best via the emotion of visceral horror. Massumi (2002) states that "viscerality is the perception of suspense" (61). The viscerally affecting experience of seeing a digitally created spider monster flying directly toward my virtual flesh creates emotions of fear and shock, which then multiplies back out toward my body, creating the affective response of chills, sweat, increased heartbeat, and rapid shallow breathing. In *Farpoint*, affect emerges via virtuality most effectively by way of targeted emotion. The emotion of fear, which is crafted from a system of conscious filtering and cultural semiotics, is a virtual response that becomes actualized in the body. It is quantifiable by its material consequences. This follows Massumi where:

> An *emotion* or *feeling*, a recognized affect, is an identified intensity as reinjected into stimulus-response paths, into action-reaction circuits of infolding and externalization—in short, into subject-object relations. Emotion is a contamination of empirical space by affect, which belongs to the body without and image. (61, italics in the original)

The emotion-based construction of immersion in VR is powerful, but in the long run, it is less compelling, as it represents little more than a trick of technology, like sitting in front of a film or even television. The impact of emotional fear works the first fifteen times or so, but as the shock factor wears off, I am less compelled to have a deep response. Shock induces fear the first couple of times, but eventually, I become accustomed to its presence and am no longer confronted by its affects in the same manner.

Instance: Immersive Encounters – Engulfing the Spectator in Sights, Sounds, and Spaces

What happens when the immersion attempts to separate itself from the act of spectating as a mode of visual stimulation? In the previous examples, the focus was on the creation of immersion in virtual worlds via televisual representation paired with embodied actions. In Complicite's 2016 production of *The Encounter*, the perceptual apparatus of mediatized spectators activates using sound. In his monograph on how sound is used to gain the attention of theatrical spectators, George Home-Cook (2015) argues that "listening as a species of attention entails an embodied movement in and through environmental space" (139). He further argues that the perception of sound has the capacity to engage spectators in an enhanced "sense of being-present" where being "thus conceived, is immanent, self-present, all-encompassing, and 'immersed'" (139). While the primary concern of immersivity via VR is often visual stimulation, aural attention is a sensory capacity that we should not overlook as a powerful component in the creation of virtual worlds. In *The Encounter*, sound and the listening spectator enter into a dynamic tension where imagination and perception work together to fully engage in an experiential narrative.

I first experienced *The Encounter* at the Brighton Festival in May of 2016 and then again on the closing weekend of its Broadway run in January of 2017. The show is a monologue-based meditation on time and consciousness told by Simon McBurney. While the production's primary narrative concerns the photo-journalist Loren McIntyre's journey into the Amazon rainforest and his encounter with the real-life indigenous people, the Mayoruna, the backdrop is technology's impact on time and memory. The show begins with McBurney discussing the smartphone and its ability to supplant memory as a digital device that augments people's minds and offloads all our cherished moments. This line of thought also partially argues for a loss of imagination due to the adoption of technologies of virtuality. As a way of contesting this loss, McBurney uses technology to reengage an ability to mentally project space, place, time, and story using binaural audio and digitally manipulated Foley sound effects.

For much of the production, the narrator delivers his story in a simple and calm, nearly whispered voice which is picked up by a binaural microphone. As they are seated, each audience member is given a set of specially linked Sennheiser HP 02–100 headphones that deliver the audio piped directly from onstage. At one point, I took off my headset to check on the actor's volume and found that sitting seventy feet away, I could hear nothing. Onstage is a dummy-head mic used to record atmospheric sound which is mixed to overlay the narration. The dummy head has microphones embedded on both sides where the ears would be replicating the effect of human hearing. By recording the audio in this fashion, sounds have the same dynamic they would as if heard in a live space. The audio is captured from all around the head as opposed to from one specific direction which gives the impression of immersive spatiality. Environmental sounds are recorded using practical foley effects, such as the crunching of a potato chip bag for the sound of boots on dried leaves, or the slosh of liquid in a water bottle for a stream. Digital feedback effects create the buzz of mosquitos. Each of these sounds is recorded live, and as McBurney shows us, are then sampled, looped, and synthesized into a jungle atmosphere. Binaural audio is utilized in the production as a tool to replicate the way humans hear. Gareth Fry, the production's sound designer, explains binaural audio as:

> … a way of replicating the human hearing system, so if you were to record something binaurally and listen to it back over headphones it's exactly like you being there and it's better than any other technology, including surround sound, at recording a space … it's a great technology for telling stories and taking you to unusual places like the Amazon rainforest and it places you, as the audience, on stage with the performer. It creates this wonderful sense of intimacy.
>
> (Savage 2016)

As the aural soundscape develops, I begin to experience the sense of being at the center of the narrative. I have gained the ability to experience the work from performer McBurney's perspective. Sound is no longer being delivered to me from the stage, but rather, I am on stage virtually, living the experience told by McBurney. The audio immerses the audience in the narrative, and since the story is told instead of presented, it requires the spectators to engage in tricks of imagination to create the visual representation of the Amazon in their own mental plane.

In most examples of theatrical immersion, there is an expectation that the spectator is thrown into some form of elaborate scenographic space where they are led to believe that they are somehow part of the narrative world. In *The Encounter*, spectators rely on the relationship between the sensory manipulation of sound and the imagination to create meaning. Home-Cook (2015) argues, "imagination is not restricted to the act of looking nor to the realm of the mind, but rather consists of a complex and continuous interplay among the senses of touch, audition, vision and proprioception" (142). One does this while seated in the traditional proscenium audience/stage configuration. This form of immersion shows how spectatorship is not just a passive function based on a visual relationship. Alston (2016) explains, "affect then implicates the audience not just as a judgmental and potentially empathetic observer of a fictive world and its inhabitants, but as an essential part and co-producer of that world" (36). The spectator of *The Encounter* is implicit in the action via their own mental projections developed through their affective senses. Using their connection to the technology, the audience imaginatively creates the narrative world they will move through. The spectator creates the sense of immersion virtually in their mind through an auditory mode of sensory engagement.

During my first visit to the production, I traveled down to Brighton from London on little rest and admittedly began to fall asleep during the production. In doing so, I began to better understand how seeing was not a necessary element of immersive spectatorship. I participated in the production via other sensory aspects of my perceptual apparatus. This revelation further ingrained in me the idea that mediatized spectatorship is more than a unidirectional reception of information from a stage spectacle, but rather, a relational operation between multiple elements of time, space, embodiment, and consciousness. This aligns with a mediatized conception of being-in-the-world; one where technological tools help to construct the reality we perceive. Similarly, when McBurney interviewed members of the Mayoruna while developing the narrative, he "asked them where they thought consciousness lay?" Their answer is not "behind the frontal lobes. Instead, they pointed to the forest. For them, the exterior world and interior worlds are interconnected" (Marcus 2016). Again, the overlap between virtuality and actuality becomes visible, but in this case, virtuality develops in the mind of the spectator via the material soundwave they receive through the headphones. Sound manipulation instigates an interaction with the spectator's mind, creating the virtual world of the Amazonian jungle. To immerse oneself in this virtual world of imagination seems most easily accomplished by closing one's eyes and listening deeply. This made me wonder how much immersion is predicated by a construction in the mind, and that space, time, place, and other aspects of reality are simply representations developed mentally but accessed by visceral and affective manipulation. The interaction between the perceptual apparatus of spectators and the spectacle in the many different representational configurations is what allows the mediatized spectator to transport itself into an immersive environment.

Instance: Bridging Divides – Agency and Affect in Corporeal and Virtual Immersion

The perception of virtuality via immersion and its relationship to embodied experience through digitally matrixed interaction is a major part of the mediatized condition in experiential spectatorship. Digital interfaces simultaneously allow broader and deeper connections with(in) performance by expanding the ontology of location/space, while at the same time narrowing what it means to be in proximity to people, places, and things. In the following, I explain how differences between technological/digital immersion and corporeal immersion help to tease apart the agential possibilities of virtuality.

In May of 2012, the MIT Media Lab conducted a trial experiment on the production of *Sleep No More* in coordination with Punchdrunk. This extended experience operated by tethering an on-site (actual) immersant wearing digital sensors to an off-site (virtual) immersant experientially interacting with both the digitally captured immersive environment and the on-site immersant via a computer monitor and keyboard. The experiment tested a virtually accessed version of immersion to create a unique experience where "online participants partner with live audience members to explore the interactive, immersive show together" (Torpey et al. 2012). Together is a bit misleading considering the virtual immersant is the only one fully aware of their connection to the actual immersant. The NY Times reviewer David Itzkoff was told that he "was bound together" with an unknown Other by accepting the challenge to help "a troubled ghost" after communicating via Ouija board with the virtual immersant (Itzkoff 2012). According to Punchdrunk's director of enrichment, Peter Higgin (2012), the challenge presented was "to recreate the infinite possibilities for journeys and experiences happening simultaneously across a Punchdrunk production." By analyzing this tandem experience's operation, I intend to show how differing modes of agency change the actual immersant's and virtual immersant's sense of being a necessary part of the event.

The Media Lab mediated the immersive experience to test the range of possibilities in augmenting live performance using digital virtuality via telematics and haptic feedback. The goal was to understand further the ways immersive experiences can be transplanted via virtual systems into an at-home experience that allows its spectator the ability to negotiate real-world physical space via a live avatar in the online platform. The MIT project used the digital augmentation to foreground interaction between its users, giving them unique modes of experiencing both the environment and agency. For the virtual immersant, mediation activates the perception of digitally augmented embodiment lived daily in the twenty-first century with the expectation that "the more intimately manipulated the technology the more embodied the experience is perceived to be" (Machon 2013, 36). Unlike the traditional *Sleep No More* experience, both spectators' narrative exploration was closely monitored and tailored for modes of interaction. A customized mask with haptic transducers and environmental RFID trigger sensors that control physical elements in the space modified the actual immersant's experience. The virtual immersant followed the actual immersant via the web interface using text-based commands while having video and audio streamed over digital equipment to induce a sense of immersion at home. The goal was to engage the virtual immersant's perceptual apparatus via technological mediation.

Where the immersive framing of the traditional ambulatory experience in *Sleep No More* allows proximity between the actual immersant and curated locations through the choice of leisurely navigation, the MIT mediation bonds the actual immersant to the virtual immersant, creating a sense of tension in the experience of agency. The virtual immersant accesses proximity to the event and space digitally, giving them the ability to guide and alter the experience for the actual immersant using interactive tools. For example, the sensors in the mask worn by the actual immersant create a proprioceptive link, and the inclusion of remotely operated typewriters allows the virtual immersant to communicate to the actual immersant like a spectral guide. By manipulating the actual immersant in the physical space using text prompts, the virtual immersant achieves a form of extended agency. Like playing a video game, the virtual immersant gains a heightened level of creativity by manipulating the actual immersant's journey through the digital feedback as if the actual immersant is a computer simulated avatar. Measurable agency is gained through the manipulation of the actual immersant in the physical space. The variety of sensors attached also delivered multiple forms of biofeedback for the MIT researchers. The operators used this feedback to quantify the subjective experience of the actual immersant. One question the linking of virtuality and actuality presents is: What mode of agency did these mediatized spectators encounter: affective or tangible?

Affective agency is the embodied feeling of agency, while tangible agency is measurable in an empirical context by the appearance of a definable change based on the spectator's input. The former mode of agency is only consequential to the individual in the event and has little bearing on the overall operation of the event. The experience focuses more on personal affective response than a tangible agency to concretely alter the world immersed in, and for this reason, I consider it *affective agency: A mode which produces the illusion or feeling of participation*. Affective agency exists in the virtual factory of the mind but manifests relative to the actual environment through the body. The experiential nature of the affective response intensifies when one is immersed inside meticulously planned scenographies, allowing interactions between designed space and sensorial feedback, but the outcome of the event is still dictated by the makers and often inflexible. However, the immersant experiences a sensorial impression (feeling) of agency to impact the outcome. This feeling of agency allows an affective response arising from differential meaning-making and sensory input. Like the exposure to fright in the *Farpoint* example, an immersive experience that relies on affective agency tends to lose its impact upon multiple exposures. This is why there is often a necessity to couple the immersive experience with other architectures for it to have lasting resonance.

Punchdrunk and MIT ended the collaboration after a five-day testing process. The experiment showed differing modes of agency among the two subject positions. Through the digital mediation, the actual immersant became a corporeal avatar, whose experience was controlled by the virtual immersant. When digitally tethered, a new level of authorship grafts onto the experience for the spectator in the event, delimiting choice and taking away their perceived agency to create individualistic narrative meaning and subsequently the affective experience. The experience for the virtual immersant is different due to the ability to control the other spectator, exuding tangible agency through the outcomes of interaction with the scenic/narrative environment via the technology. For the virtual immersant, there was also a loss because the technological manipulation could not match the affective impact of being in the corporeal space. What differentiates the two concerning experience is the difference between who feels the most agency and who has the most agency.

The experiment attempted to navigate the subtle difference between immersion and participation by navigating both at once. This attempt replicated the duality and the overlaps between virtuality and actuality. By tethering experience via the two separated spaces—one physical, the other virtual—a tangibly heightened sense of agency through the alienation of the two separate interfaces arose. For the immersant in the physical space, the sensorial interaction via haptic, aural, and visual stimuli allowed a subconscious effect to guide expectations about individual experience. These expectations controlled the immersant through their journey of the curated space. The actual immersant had the agency to observe but had little control over the outcomes of the event, even though they were lured into a feeling of control through their own experiencing of the event. The virtual immersant gained a higher level of measurable agency through the tangible outcomes of their interaction with the event and the actual immersant experiencing the space. By typing into the keyboard directions or thoughts, which were responded to in the physical space, there was a direct correlation between action and event, as opposed to a simulated affective response through sensorial interfaces inside the event.

The virtual immersant could tangibly impact the experience of the event in ways more measurable than strictly via personalized affect. The technology allowed agency that is both creative and tangible, affecting not only the participant's experiential feeling of the event but also the event itself. The virtual immersant's input created a feedback loop that effected and affected the actual immersant while also changing the exploration of the narrative for both. In Biggin's (2017) account of this experiment, she argues that the technology transformed the immersive experience into a "ludic one" or game (167). As such, the narrative elements of the experience were subjected to a new form of experiential manipulation, negating some of the immersive affect for both the virtual and actual immersant. Considering the purpose was never to create a game-based experience, it is understandable that the experiment ended without the technology being implemented further. It might be that the agency gained through the technological interface may have been simply too much for Punchdrunk's designers. It possibly negated their supposed gift of open-ended meaning-making for the actual immersant. As Higgin (2012) explains, one of the difficulties with the project was how they "were treading a fine line between game and experience, in an already delicately balanced performance." His statement marks the necessity of understanding immersion as both an aesthetic and an architecture for experience that often operates as a form of interaction coupled with other architectures.

Conclusion: Ethical Conundrums of Traversing Immersive Realities

I first encountered *Sleep No More* as an audience member in the winter of 2012. I was disappointed in some ways, yet incredibly intrigued by the affective experience of the production. This was the first theatrical event that allowed me to have this much freedom as I was invited to explore to my heart's content. Exploring in this manner is a form of agency that Machon explains involves "processual interaction through the event, [and] shapes the unique journey for each participating individual" (Machon 2016a, 36). The body of the spectator interacts with the narrative world in the event, and by doing so, it gains a sense of agency that "shapes and transforms potential outcomes of the event" (36). The experience creates a false understanding of impact induced by the ability to direct your own attention toward whichever avenue in the story-scape chosen. Unfortunately, there is only one potential outcome of the event as dictated by the makers. A feeling of agency exists, allowing one to think they can impact the outcome, and this allows the immersant in the event an affective response leading to differential meaning-making, which gives the impression that they can impact the structure of the event. As the elevator operator explains when venturing into the space, "fortune favors the bold"; an invite to perform in an active manner is at the heart of the performance, but this invitation is a conceit that can have mixed results.

As a spectator in this form of immersion, the lack of defined choices, clear rules, objectives, or personal identity diminishes the ability to pin down any form of specificity within the experience beyond feeling. Aesthetic meaning-making forms the basis of the immersive experience based solely on one's subjective and individualized sense of being in the immersive world. Primary agency comes from the way in which one consumes or interprets the narrative through this bodily affect. This corresponds to Gareth White's (2013) explanation of immersion: "Not all audience participation would be claimed under the rubric of immersive … To be inside the work, not just inside its physical and temporal space but inside it as an aesthetic, affective, phenomenological entity gives a different aspect to the idea of a point of view, and of action" (17). It is this phenomenologically modified point of view that I am marking as the foundational aspect of immersion in the model presented by Punchdrunk and in many VR experiences. This model is also the form of mediatized spectatorship that I would like to critique because of its connection to perceived agency as an affective response with little resonance beyond an impact on the emotions of the individual. Even though one feels the immersive experience so strongly while inside the event, it tends to be contained and significant only on an individualistic basis. Alston (2016) is also critical of these experiences and connects them to a culture based in the experience economy and neoliberalism that relies on narcissistic participation for contemporary hegemonic structures to operate. Alston states:

> Aesthetic experiences in immersive theatre tend to promote introspection, because in the heady heights of immersion and participation it is not art objects that take precedence so much as the affective consequences of an audience's own engagement in seeking, finding, unearthing, touching, liaising, communicating, exchanging, stumbling, meandering and so on, each geared toward the promotion of peculiarly intense or profound experiences… (7)

The experiences that the seeker of immersion expects are those that allow them the feeling of control over their affective states and a sense of agency in aesthetic meaning-making even when no control exists.

At the early takeoff of scholarship on immersive theatre in 2014, *Theatre Journal* devoted an entire issue to spectatorship in contemporary theatre. In an article from that issue, Karen Zaiontz (2014) discusses the nature of "narcissistic spectatorship" that is inherent in the practices of immersive theatre: "A narcissistic spectatorship encourages the viewer to fully engross herself in an artistic production in a way that highlights her own singular relationship to the piece. The spectator is not positioned as an author or agent who has the power to create or enact concrete change, but as an *experiencer* of the piece" (408–409). The nature of experience for this form of spectator has developed away from that of audience as a formal group and, instead, focuses on the individual who has become accustomed to crafting his own self using contemporary technological tools. In many immersive media, there is a propensity toward individualism that is inherent in the neoliberal condition where capitalism encourages a focus on consumption that is increasingly focused on the individual. This is one of the negative repercussions of the fully enveloping nature of virtuality in contemporary society. Some immersive theatre practices might possibly be subverting the early potential of virtuality that promised an escape and possibility to upend the strictures of late twentieth-century capitalism. Though immersive practices are praised for their ability to bring in a younger demographic of theatre audiences, Alston (2016) warns, "immersive theatre nonetheless risks serving neoliberal capitalism by fetishising the co-opted feeling body and celebrating a potentially profitable, individualistic and apparently personal form of consumer productivity" (158). Attempting to connect what has been historically looked at as a community event (theatre), to the needs, wants, and desires of the individual consumer is an example of the shift toward a co-option of the mediatized spectator.

The popularity among non-traditional theatre attendees of projects like *Sleep No More* has caused Punchdrunk to be a primary exemplar within scholarship on immersion. The commercial success of the production allows it to extend into a perpetually running event with three operating hospitality attachments in multiple cities. When the production first contracted inside the McKittrick hotel, a "smoky" speakeasy named the Manderlay Bar was included as part of the 2011 run. In 2013, a rooftop bar, the Gallow Green, and restaurant/music venue, The Heath, were added to the building, accessible even without a ticket to the performance. These hospitality tie-ins have become secondary narrative venues augmenting the reality of the immersive experience. Punchdrunk and their experiential design partner, Emursive, blur the boundaries between fictive experiences and reality through an invasive enveloping of multiple worlds via consumerism when partnering with companies such as Sony, Louis Vuitton, Heineken, and Stella Artois. The inclusion of the expansion into commodity culture based on individual consumption, as seen through these partnerships, is one of the negative aspects of the venture into immersive narratives.

Immersivity has also unfortunately become a catchphrase used to attract consumers who want the appearance of ultimate control over their daily experiences. You can see this in multiple forms of marketing today. Alston devotes the entire chapter to the development of a culture of narcissistic immersion that is inherent in the makeup of immersive events. His critique focuses on the invitation to engage with(in) the event as a way of accessing individualized affective experiences impacted by the interface between space and narrative. This affective experience "implicates the audience not just as a judgmental and potentially empathetic observer of a fictive world and its inhabitants, but as an essential part and co-producer of that world" (Alston 2016, 36). The effect of this experience is only a form of faux-participation because it has no consequences beyond the individualized feeling-self. Compare this situation to the differences marked between immersion and participation in the following chapter. As an immersant, a spectator often becomes part *of* the world but does not materially take part *in* the world. Their beingness in the event has little impact beyond narcissistic feeling. With the recent closure of the New York City location of *Sleep No More* in 2024, I wonder if that feeling is running out.

Notes

1 I avoid non-western conceptualizations of reality that might arise from culturally specific foundations of spirituality and essence, as it unnecessarily complicates the relationship between virtuality and actuality as the basis of technogenetic change covered in this project. For a wider exploration of the multi-cultural concepts of reality, I suggest David Chalmer's *Reality* + (2022), which is referenced here.
2 For examples of artists using VR technologies, see (Dixon 2007:367–394; Birringer 2006; and Giannachi 2004).
3 An archived trailer for the experience with representations of the technology is available here: https://www.youtube.com/watch?time_continue=32&v=_QIbI4Wtgug

Bibliography

Alston, Adam. 2016. *Beyond Immersive Theatre: Aesthetics, Politics and Productive Participation*. Palgrave.
Bay-Cheng, Sarah, Chiel Kattenbelt, Andy Lavender, and Robin Nelson. 2010. *Mapping Intermediality in Performance*. Amsterdam University Press.
Baudrillard, Jean. 1995a. *Simulacra and Simulation*. Translated by Shelia Glaser. University of Michigan Press.
Baudrillard, Jean. 1995b. *The Gulf War Did Not Take Place*. Translated by Paul Patton. Indiana University Press.
Baudrillard, Jean. 2014. *Screened Out*. Translated by Chris Turner. Verso.
Benford, Steve, and Gabriella Giannachi. 2011. *Performing Mixed Reality*. MIT Press.
Biggin, Rose. 2017. *Immersive Theatre and Audience Experience: Space, Game and Story in the Work of Punchdrunk*. Palgrave.
Birringer, Johannes. 2006. "Digital Performance." *Performance Research* 11, no 3: 42–45.
Bolter, Jay David, and Richard Grusin. 1999. *Remediation: Understanding New Media*. MIT Press.
Carr, Nicholas G. 2010. *The Shallows: What the Internet Is Doing to Our Brains*. W.W. Norton.
Causey, Matthew. 2006. *Theatre and Performance in Digital Culture: From Simulation to Embeddedness*. Routledge.
Causey, Matthew. 2016. "Postdigital Performance." *Theatre Journal* 68, no 3: 427–41.
Chalmers, David. 2022. *Reality +*. W.W. Norton & Company.
Deleuze, Gilles, and Felix Guattari. 1987. *A Thousand Plateaus: Capitalism and Schizophrenia*. Translated by Brian Massumi. University of Minnesota Press.
Dixon, Steve. 2007. *Digital Performance: A History of New Media in Theater, Dance, Performance Art, and Installation*. MIT Press.

Foster, Thomas. 2009. "Virtuality." 317–28. In *The Routledge Companion to Science Fiction*, edited by Mark Bould, Andrew Butler, Adam Roberts, and Sheryl Vint. Routledge.

Frieze, James. 2016. *Reframing Immersive Theatre: The Politics and Pragmatics of Participatory Performance*. Palgrave.

Giannachi, Gabriella. 2004. *Virtual Theatres: An Introduction*. Routledge.

Gibson, William. 1984. *Neuromancer*. Ace Publishing.

Greene, Andy. 2015. "Flashback: 'The Wizard' Unveils Nintendo's Power Glove - Rolling Stone." *Rollingstone.com*. http://www.rollingstone.com/culture/videos/flashback-the-wizard-unveils-nintendos-power-glove-20150528

Haraway, Donna. 1991. "A Cyborg Manifesto: Science Technology, and Socialist-Feminism in the Late Twentieth Century," 149–81. In *Simians, Cyborgs and Women: The Reinvention of Nature*. Routledge.

Hayles, N. Katherine. 1997. "The Condition of Virtuality." 183–208. In *Language Machines: Technologies of Literary and Cultural Production*, edited by Jeffery Masten, Peter Stallybrass, and Nancy Vickers. Routledge.

Hayles, N. Katherine. 1999. *How We Became Posthuman: Virtual Bodies in Cybernetics, Literature, and Informatics*. University of Chicago Press.

Higgin, Peter. 2012. "Innovation in Arts and Culture #4: Punchdrunk - Sleep No More." The Guardian. London. May 25, 2012. https://www.theguardian.com/culture-professionals-network/culture-professionals-blog/2012/may/25/punchdrunk-digital-innovation-sleep-no-more

Holland, Todd. 1989. *The Wizard*. USA: Universal Pictures.

Home-Cook, George. 2015. *Theatre and Aural Attention*. Palgrave.

Immersive. 2024. *Dictionary.cambridge.org* https://dictionary.cambridge.org/us/dictionary/english/immersive

Immersive. 2024. *Merriam_Webster.com*. https://www.merriam-webster.com/dictionary/immersive

Immersive. 2024. *Oxfordlearnersdictionaries.com*.

Itzkoff, Dave. 2012. "A Guinea Pig's Night at the Theater." *New York Times*. May 23, 2012. https://www.nytimes.com/2012/05/23/theater/sleep-no-more-enhanced-by-mit-media-lab.html?_r=0

Jarvis, Liam. 2019. *Immersive Embodiment: Theatres of Mislocalized Sensation*. Palgrave.

Kattenbelt, Chiel, and Fredda Chappel. 2006. *Intermediality in Theatre and Performance*. Rodopi.

Kelly, Kevin. 2016. *The Inevitable: Understanding the 12 Technological Forces That Will Shape Our Future*. Viking Press.

Kershaw, Baz. 1999. *The Radical in Performance: Between Brecht and Baudrillard*. Routledge.

Klich, Rosemary, and Edward Scheer. 2012. *Multimedia Performance*. Routledge.

Laurel, Brenda. 1991. *Computers as Theater*. Addison Wesley Publishing.

Lavender, Andy. 2016. *Performance in the Twenty-First Century: Theatres of Engagement*. Routledge.

Lévy, Pierre. 2001. *Cyberculture*. Translated by Robert Bononno. University of Minnesota Press.

Machon, Josephine. 2009. *(Syn)Aesthetics: Redefining Visceral Performance*. Palgrave.

Machon, Josephine. 2013. *Immersive Theatres: Intimacy and Immediacy in Contemporary Performance*. Palgrave.

Machon, Josephine. 2016a. "On Being Immersed: The Pleasure of Being: Washing, Feeding, Holding." 29–41. In *Reframing Immersive Theatre: The Politics and Pragmatics of Participatory Performance*, edited by Frieze, James. Palgrave.

Machon, Josephine. 2016b. "Watching, Attending, Sense-Making: Spectatorship in Immersive Theatres." *Journal of Contemporary Drama in English* 4, no 1: 34–48.

Marcus, Joan. 2016. "Broadway's Rumble in the Jungle: How Simon McBurney Made 'The Encounter.'" *Thedailybeast.com*. https://www.thedailybeast.com/broadways-rumble-in-the-jungle-how-simon-mcburney-made-the-encounter

Massumi, Brian. 2002. *Parables for the Virtual: Movement, Affect, Sensation*. Duke University Press.

Munster, Anna. 2006. *Materializing New Media: Embodiment in Information Aesthetics*. Dartmouth College Press.

Murray, Janet H. 2017. *Hamlet on the Holodeck*. 2nd ed. Free Press.

Packer, Randall and Ken Jordan. 2001. *Multimedia: From Wagner to Virtual Reality*. Norton.

Parker-Starbuck, Jennifer. 2011. *Cyborg Theatre: Corporeal/Technological Intersections in Multimedia Performance*. Palgrave.

Ryan, Marie-Laure. 2001. *Narrative as Virtual Reality: Immersion and Interactivity in Literature and Electronic Media*. Johns Hopkins University Press.

Savage, Adam. 2016. "Complicite's *the Encounter* - Bringing Binaural to the Barbican." *Audiomediainternational.com*. http://www.audiomediainternational.com/live/complicite-s-the-encounter-bringing-binaural-to-the-barbican/05265

Singer, Alexander, dir. 1993. *Star Trek the Next Generation*. Season 6, Episode 12, "Ship in a Bottle." Paramount Television.

Tach, Dave. 2016. "Farpoint Is the VR Shooter I Wanted, but I Didn't Want the VR Sweats." *Polygon.com*. https://www.polygon.com/features/2016/7/8/12118898/farpoint-is-the-vr-shooter-i-wanted-but-i-didn-t-want-the-vr-sweats

The Void. 2016. "Hyper-Reality Is Taking Virtual Reality to the next Level." *Blog.thevoid.com*. https://blog.thevoid.com/how-hyper-reality-takes-virtual-reality-to-the-next-level/

Torpey, Peter, Ben Bloomberg, Elena Jessop, and Akita van Troyer. 2012. "Remote Theatrical Immersion: Extending Sleep No More." *MIT Media Lab*. October 21. www.media.mit.edu/posts/remote-theatrical-immersion-extending-sleep-no-more/

Virtual. 2017. "Virtual, Adj." *OED Online*. Oxford University Press.

Wachowski, Lana, and Lilly Wachowski. 1999. *The Matrix*. United States: Warner Bros.

Westling, Carina. 2020. *Immersion and Participation in Punchdrunk's Theatrical Worlds*. Methuen Drama.

White, Gareth. 2012. "On Immersive Theatre." *Theatre Research International* 37, no 3: 221–35.

White, Gareth. 2013. *Audience Participation in Theatre: Aesthetics of the Invitation*. Palgrave.

Zaiontz, Keren. 2014. "Narcissistic Spectatorship in Immersive and One-on-One Performance." *Theatre Journal* 66, no 3: 405–35.

3　The Architecture of Participation

3.0　The Promise of Web 2.0: Participation, Convergence, and Collaborative Technologies

Think back to your earliest memories. When you were young, there was an endless number of possibilities to choose from. These choices presented themselves in every moment of your life, and they allowed a variety of experiences to unfold before you. How did you manage these choices? Did you think about what would be individually fulfilling in the moment, or did you think about the possible long-term effects of your choice? Were you conscious of the impact that your choice might have on others or on the world around you? Did you include others in your process of choosing? Was there a time when your daily interactions seemed to be full of possibility and that the digital tools you used seemed to offer an almost utopian sense of connections and possibility? What about now? How have those opportunities to make a choice with real consequences changed? In what ways are the choices you make actual choices? Are your choices real, or are they just a mirage presented as a way of keeping you on a specific track? In an age of deep mediatization, what is at stake if the politics of direct participation are largely void of meaning?

Throughout this chapter, I argue that a participatory social condition prompted by Web 2.0 resources can be utilized to better understand the modes of exchange activated by a participant spectator in practices that rely on tangible agency via direct action and choice making. Web 2.0 tools emerging in the late 1990s and becoming dominant media logics in platforms of the late 2000s created a paradigm through which mediatized spectators are often conditioned to seek material participation through the technic of *networked collaboration*. Under this technic, the primary mediator of human communication has become the digital sphere of the internet and the many individualized domains contained in that sphere. I refer to these domains as the unique spaces we call social media, such as YouTube, Twitter (now X), Facebook, Reddit, and the now seemingly lost Myspace. Each has its own logic and purpose but operates based on the necessity of the creation of content and the distribution of discourse between participating spectators. The prevailing

DOI: 10.4324/9781003398813-4

mode of interaction with these domains requires active engagement on the part of the participant as a collaborator with both the domain and the other users who recirculate the effects and affects of the content created with(in) the domain. This engagement operates as a form of collaboration between a network of interconnected and participating media users.

In networked collaboration, mediatized spectators gain a visible ability to make tangible change and impact in the action and dramaturgy of their everyday lives and likewise their performance events. Web 2.0, or what has been described as the participatory web (Jenkins 2006), enacts a technogenetic process on its spectators that encourages a desire for reciprocity and communal action between multiple agents in networks of sociality. The technological spaces inhabiting the digital become real by constant interaction between their spectators, as creators and remixers of social configurations. Understanding how spectators materially participate with technological domains, as interlocutors of being and selfhood, aids in further developing the analytical frame for spectatorship with(in) paradigms of mediatized sociality. A foundational aesthetic of the *Architecture of Participation* is how it allows participants *tangible agency*, where the effects and affects of deliberative and democratic choice making, alongside direct and material collaboration, create change in both the event/space and beyond the event/space.

Participation is covered to explain how Web 2.0, and specifically platforms and domains for social media, have the capacity to promote ethical and communal exchange as an aspect of experiential spectatorship. The experience of the spectator with(in) this architecture is predicated on activating their ability to serve in a democratic fashion through discourse, decision making, and direct action. This material form of participation is different than other forms of experience in its capacity to form a direct effect beyond the spectator. While still active, watching in the so-called "passive mode" is an endeavor that does not have the same consequences as material participation because it operates through an individualized form of engagement with biological and contemplative processes found in the mind and body of the perceptual apparatus. While the act of theatrical spectatorship is often referred to and contested as a passive event, Jade Rosina McCutcheon (2013) explains, "watching an actor onstage is more than a visual event; theatre is a transaction, a sharing of ideas and a site of reflection" (147). This is different by degrees from purposeful acts of participation within the event, though. McCutcheon refers to the process by which we gain knowledge and embody that same knowledge in the biological/contemplative process of watching that engages our mirror neurons and cognitive approaches to identification via mimesis (Cook 2008; McCutcheon and Sellers-Young 2013). This element of the brain processes what we see and "is thought to be responsible for action understanding, intention, emotional attunement, communication, joint action and imitation," (Cook 2008, 590). Activating the mirror neuron does not require any form of active material engagement with the event beyond mental processing connected to the visual registers of the perceptual apparatus. Via this neuronal activation, spectators might engage in an embodied cognition of that which they perceive, but there is no material impact beyond the self. Engaging the mirror neuron simply allows an "understanding of the self, how it is viewed by others and how we can best articulate our identity to others" (McCutcheon 2013, 146), which is impactful but only within a narrow solo positionality. Through the *Architecture of Participation*, the entirety of the event has the capacity to change based on the spectator's input and therefore has increased impact for both the individual and those within the community of the event, offering a performance-based mode of relational discourse. This discourse works to rewire the dramatic function of reception one might experience in a media event where the action "on stage" is separated from the action of the audience. Due to this, it is fruitful to consider how participation capitalizes on the experiential capacities of postdramatic aesthetics.

Hans-Thies Lehmann (2006) states, "It is a fundamental fact of today's Western societies that all human experiences (life, eroticism, happiness, recognition) are tied to *commodities* or more precisely their consumption and possession (and not to a discourse)" (183). Lehmann offers this challenge as a way of defending the shift from dramatic form to postdramatic form within the confines of theatre and performance. He sees that the shift to this form is partially wrapped up in raising consciousness toward what he describes as a "politics of perception" (185) through which citizens become spectators in their ongoing relationship to the media sphere. His argument here is built off Guy Debord's (1995[1967]) theories that presciently called out the societal influence of the media and culture industries in the late 1960s. Debord's *Society of The Spectacle* is relevant to introduce in relation to this chapter as it frames the spectator as one who has been captured by the forces of capitalism causing their political power to be drained, leaving them prisoners of the image, a metaphor for the glossy sheen that culture is depicted in through film and television. Media for Debord was representational and therefore could never allow anything other than unidirectional transfer of information. He made this argument as a way of waking the spectator up from their stupor in order to reactivate their political agency through direct participation in the societal challenges they were facing. In reading Debord as a precursor to the political activist bent toward art in the 1960s, Claire Bishop (2012) highlights the heart of his argument:

> It rehumanises a society rendered numb and fragmented by the repressive instrumentality of capitalist production. Given the markets near total saturation of our image repertoire, so the argument goes, artistic practice can no longer revolve around the construction of objects to be consumed by a passive bystander. Instead, there must be an art of action, interfacing with reality, taking steps—however small—to repair the social bond. (11)

I offer Debord and Lehmann here as a way of questioning the ways a participatory mode of spectatorship might challenge a notion that contemporary culture has divested from community, focusing instead on individualism.

In direct confrontation to the tranquilizing effect of both theatrical mimesis and televisual media, the Architecture of Participation supports modes of tangible agency through networked collaboration that is found both in the technologies of Web 2.0 and in the collective agency of participating spectators. The emphasis is on the relationship between choice making and collaboration as a way of activating political energy. Participant audiences operate in modes that require give and take between various nodal agents—both human and technological— to allow impact within the event and the narrative throughline of the event. If immersion is largely an individualized experience relying on affect, through participation, the spectator/media network works together in some fashion toward a common though flexible endpoint. This participant audience network allows the emergence of a political assemblage activating democratic potential via the choices the audience makes and the collaboration they engage within. I refer to Lehmann (2006) as a nod toward the format that participatory events are often framed. The aesthetic frame of the postdramatic often contains these events due to their lack of emphasis on a dramatic and static dramaturgical structure of internal dialogue, or defined linearity, instead relying on the influence of a participant as sculptor of action and intent through relational actions and direct choice making. Lehman states, "It is not through the direct thematization of the political that theatre becomes political but through the implicit substance and critical value of its *mode of representation*" (178, italics in original). Participation operates as a mode beyond simple representation to incorporate the mediatized spectator in an interchange where the agency engaged causes a form of action and exchange similar to Lehmann's political aesthetics. In the Instances covered in the hinge section of this chapter, a tangible form of agency and political efficacy is accessible through a spectator's transformation into a participant who acts as the direct conduit of democratic energy through communication and/or material participation. These Instances rely on modes of exchange that could not occur without the visible input of the participating spectator. They exemplify how participants and participatory performance practices develop notions of relationality that allow a form of mediatized politics to emerge.

The Participatory Condition as Technology: Networked Collaboration and Social Worlds

Recalling Couldry and Hepp's (2017) model of communication under deep mediatization, information passes between humans and other humans—and likewise machines—in mediated platforms, or the nodes we call social media, and these platforms themselves construct the very idea of a social sphere through the relations they set up. They explain,

> The social world is the intersubjective sphere of the social relations that we as human beings experience. Those relations are rooted in everyday reality, a reality nowadays always interwoven with media to some degree. The social world is, in turn, differentiated into many domains of meaning, even though it is bound together by multiple relations of interdependence and constraint. (20)

All media that are socially manipulated operate as conduits or threads connecting humans and other objects with agency via various forms of communication. The threads can be interwoven into ethical and communal configurations via the technic of networked collaboration. As part of a mediatized mode of perception and action, this form of collaboration can graft onto acts of spectatorship that "emancipate" the spectator from its chains of liberal individuality to better understand ways of operating with(in) democratic assemblages. The logics and operations of social media, which are part of what has been described as Web 2.0 (DiNucci 1999), are a part of the overarching mediatized social world that helps inform the mode of perception covered in this chapter. This model for perception is part of what performance and media studies scholars (Barney et al. 2016; Jenkins 2006) call the participatory condition; an embedding of the earlier conceptualization of participatory culture, first introduced as a socio-cultural framing by Henry Jenkins in the early 1990s.

Like the condition of virtuality discussed in the previous chapter, the participatory condition develops through technogenetic means through the linkages established via the many entanglements found in the networked foundations of Web 2.0. Through networked collaboration, people and subsequently spectators are programmed to seek out mediatized connectivity using their digital tools, bringing about a paradigm where simple receiving of information becomes less activating as a mode of exchange. Instead, active participation becomes both encouraged and necessary. A participatory condition is therefore one where interpersonal action and dialogue create new possible configurations of the social itself. The participatory condition is also a "contextual feature of everyday life" (Barney et al. 2016, vii), where all aspects of social, cultural, economic, and political activities are developed out of human participation. Because of changes brought about by Web 2.0, participation weaves the fabric of contemporary life and "becomes the measure of the quality of our social situations and interactions" (ix). This fabric informs all virtues of civic development, including "equality, justice, fairness, community, [and] or freedom" (ix). In the mid-2000s, this participatory condition started to become part of the everyday through the influence of technologies embedded within Web 2.0. The condition impacts all levels of human interaction, including spectatorship, where taking part includes both the reception and production of artistic content as a network of user-generated relational exchange.

Discussing the ongoing changes in culture(s) and social spheres at the turn of the new millennium which would spur this new condition, Manuel Castells (2010) describes networks as:

> open structures, able to expand without limits, integrating new nodes as long as they are able to communicate within the network, namely as long as they share the same communication codes (for example, values or performance goals). A network-based social structure is a highly dynamic, open system, susceptible to innovating without threatening its balance. (500)

Building off Castells to establish a theory of network aesthetics, Patrick Jagoda (2016) explains that a network at its simplest is "a structure composed of links and nodes" (8). This mode of mapping connections offers a way of spatially rendering "decentralized or distributed modes of operation" within any sphere of interactivity and/or communication (8). Jagoda utilizes networks as a techno-social framework emerging in the cultural imagination of the late twentieth and early twenty-first centuries and hardwired into social consciousness through the mass adoption of the internet. The ubiquity of the framework leads him to question the possibilities of aesthetics born through a pervasive conceptualization of interconnectivity brought about by media and communications systems. These aesthetics respond to the ways networks "suggest constant change and reconfiguration that exceeds any individual imagination and leave open possibilities" (226). They are categorized based on the myriad ways that contemporary culture works through these possibilities. He divides his aesthetics into "maximal, emergent realist, participatory, and improvisational qualities" that allow the user of various media—digital and non-digital—different access points to understanding and meaning-making (221). The goal of establishing this taxonomy is to question the way different media "prime us to undertake cognitive but also somatic and affective encounters with networks" (5). Jagoda begins with the definition above that is most concerned with how networks are understood from a mathematical and computer science perspective but also considers communications systems and the sociological and geopolitical formulations to establish his taxonomy (11). He further explains, "Networks, then, are not only theoretical figures or technological infrastructures. They also serve as organizational blueprints for different forms of economic, political, and social life" (14). No matter which valence one might utilize, the framework of the network offers an emphasis toward understanding contemporary society as one built out of a complex and ever shifting set of relations between various objects—human and non-human—with agency in flux.

Outside of these strict technological confines, a network might refer to any communication system or social organization built upon interconnection (9).

While the collective imagination largely envisions networks as part of the culture of digitalization, they are not unique to the digital. Humans have always situated themselves within network logics. They can "be composed of almost anything: computers (internet), cars (traffic), people (communities), animals (food chains), stocks (capital), statements (institutions), cultures (diasporas) and so on" (Galloway and Thacker 2007, 33). Humans—and spectators as human objects put into relation with(in) a performance system—operate as agents of relation no matter what architecture the network takes on. The logic of a networked social system is one where individuals operate as interconnected nodes that allow information to flow across and between each other in a decentralized and relational modality (Castells 2004). There is neither a fixed origin nor a stable endpoint which then necessitates a continual sense of collaboration between each node in the network. Nodes serve as central components of network ontologies, and understanding their role in network culture is imperative. The network cannot exist without the collaborative and relational nature of each node and is therefore established both through this collaboration and continues to operate through a relational flow of communication and cooperation between each node. We must remember that relation is not a stable "thing" in itself but is made up of "the relations between things" (Galloway and Thacker 2007, 37). Collaboration is in itself the relationship(s) between multiple nodes. An individual person can serve as a node in constant relationality when they see themselves as part of a network, itself a social construct made from a plethora of interrelations between multiple nodes. By adopting this logic of an individual as an informational/relational node, it is a logical leap to see how the foundation of a participant spectator is formed and guided through networked collaboration. Castells (2004, 2010) describes this collaboration as a form of cooperation based on communicative exchange, which then brings about a "networked society," a paradigm shift that aligns with the advances in the interactive computer technologies of the late 1990s and forward.

Before doubling down on the importance of participatory media as the primary promoter of network culture, we should take a quick moment to highlight a few other important influences that just happen to converge at the same historical period as the rise of the internet. The conceptualization of network culture could not have emerged if not for the advancement of the combined forces of globalization, individualization, and mediatization. Of interest is how globalization impacts a wider sense of connectivity across cultures and societies. Friedrich Krotz (2008) explains:

> ... globalization can be conceived of (a) changing of understanding of time and space and in consequence of changes in people's meaning making and actions; (b) as an increase of interactions among cultures; (c) as an increase of problems that concern all people of the world; (d) as an increasing network of transnational actors and organizations that become more and more influential both on the global economy and on political decisions. (21)

Globalization had begun largely during the rise of the colonial empires during the late Enlightenment era but most firmly took root with the deployment of media and technological processes that allowed for direct end-to-end communication across vast geographical distances. Working in tandem with the growing connectivity came a sense of increased separation from smaller formal networks of social life. As the world grew more accessible and thus possible experiences of life expanded, the notion of the individual as both a singular entity and part of a large expansive whole grew. Individuals gained the ability to see themselves as more than just neighbors to those next door, but also neighbors to those across continents and cultures. Through individualization and globalization, "traditional relations" are replaced by "abstract interdependencies and interconnections" that are best understood through the logics of the network and media connectivity (22). These metaprocess for social evolution then cojoin through media connectivity brought about by mediatization, where the multiplicity of end-to-end communication extends beyond the simple face-to-face modality only possible before the rise of global communication networks.

The emergence of the network society and likewise network sociality (Wittell 2008) has deep roots in the various aspects of mediatization but found its firm bedrock within the logic and structures of the World Wide Web and the exchange language it is structured around, namely hypertext. With the development of hypertext, the internet gained "the potential to link up everything digital from everywhere and to recombine it" (Castells 2004, 7). Like the participatory condition discussed, the networked society is based on a form of exchange between actors/nodes that situate themselves between other actors/nodes, and this creates a cultural paradigm *where process becomes more important than content.* When considering the technological capacity of the network, we must also take seriously the agency of the technological actors. The domains and informational sites are nodes with agency themselves. This aligns with Bruno Latour's (2005) conceptualization of social frames based on actor-network theory (ANT), where the network is a social configuration made up of more than just human actors who become part of the collaborative framework. For Latour, the network is also not simply a material set of objects in relation. It emerges as a symbolic force or process made of the linkages and traces of action between these linkages, for without the effort made within collaboration between nodes/actors, the network "leaves empty most of which is not connected" (133). Therefore, the entire ontology of a network is predicated on the collaborative effort of all the participants in a flattened horizon of relational interactions, and the network *as a process* survives only through these continual collaborations. For people, meaning is made not just through the primacy of the individual node/actor but in reading(s) of the relationship(s) between different nodes/actors. Latour extends the concept of network with a different valence by offering the term assemblage which Couldry and Hepp (2017) believe is helpful when considering the way "communicative practices are deeply interwoven with media technologies" (62).

They argue for combining the strength of both Castells' network society and Latour's assemblage to think of the relationship between nodes/actors as one of "*figurations* that are formed and reformed in an open-ended process" (63, italics in original). These figurations are:

> more than just association of heterogeneous elements (assemblages) and more than structures of linkages (networks). The figurations of online practice comprise an open (expanding) set of spaces for interaction and dependency, in which we are enmeshed, as we try to go doing what we do ordinarily. (65)

The open-ended and ongoing operationality of these figurations enacts a sense of relationality on the mediatized spectators embedded within. This relationality is predicated on the space between product and spectator, where collaboration operates as the site of meaning-making. Collaborative action underlies the finality—if there can be such a thing—of the participatory event.

The collaborative and participatory nature of relational exchange within these figurations also becomes the product via process. Instead of just relying on a unidirectional function of reception to deliver meaning from a pre-populated piece of content or platform to the receiving spectator, it is the ongoing sets of interactions with the content and the producers of the content that meaning is made. The content is no longer medium specific either, as meaning emerges through collaboration between multiple mediums and multiple actors within the entire system of exchange built out of interdependent figurations. I say emerges here in the sense that the essence of the product, a YouTube post for example, comes into being not simply from the posting—a form of production—but from the relational nature of the commentary from other nodes in the network that inevitably lead to a recirculation of the content either through direct "viral" redistribution or through a remediation into other domains. You see this embedded in the domains of cultural production that lead to what Henry Jenkins (2006) described as convergence culture.

Jenkins introduced a powerful argument for how early 2000s culture was predicated on a forceful sense of convergence, where the media ecosystem began to emulate the collaborative logic of the network. Convergence indicates an "ongoing process or series of intersections between different media systems, not a fixed relationship" (322). While the concept of convergence centers on the way that the media ecosphere changes to meet the needs of the network, Jenkins focuses primarily on the effect of convergence on what would have been previously understood as passive consumers. In the reception model, these consumers receive a product for their own intellectual, entertainment, and cultural consumption purposes. Instead of consumer, we should instead think of the mediatized spectator under this paradigm as a "prosumer" or one who does more than take in the product. They create/

produce the product as process through their own acts of participation and collaboration, which allows them to become collective members of a larger participatory culture. This culture and these prosumers are defined in this manner:

> The term participatory culture contrasts with older notions of passive media spectatorship. Rather than talking about media producers and consumers as occupying separate roles, we might now see them as participants who interact with each other according to a new set of rules that none of us fully understands. (3)

Jenkins was ahead of the curve in his pronouncement regarding how convergence was adopting and promoting elements of participatory culture. He had been writing about the utopian notions of digital connectivity since the early 1990s. In those early days of the consumer accessible internet, the idea of being able to instantly communicate and build a vast interconnected social sphere was thought of as a techno-dream. It was a dream that was not readily available though due to a lack of adoption and deployment of a fully integrated network infrastructure that could only emerge once more nodes were brought online. It took the creation of a second wave of networked, interactive, and participatory digital tools to be realized in a way that could fully allow participatory culture(s) to thrive on an interconnected global scale.

In later work, Jenkins is clear in explaining that all forms of social practice have inherently participatory elements (Jenkins, Ito and boyd 2016, 12). So, we should not solely attribute participatory culture to technologies and media. That said, there has been a historical shift in the ways technologies of communication allow for the rise in participatory culture, due to the rise in interactivity as a basis for media use. Technologies that adhere to the logic of Web 2.0 have a foundation based on user interactivity. Interactivity and participation are not synonymous but are interconnected.

> Interactivity refers to the properties of technologies that are designed to enable user to make meaningful choices (as in a game) or choices that may personalize the experience (as in an app). Participation, on the other hand, refers to properties of the culture, where groups collectively and individually make decisions that have an impact on their shared experiences. We participate *in* something; we interact *with* something.
>
> (11, italics in original)

What technologies associated with Web 2.0 brought about was an expansion of scale in the ways participation could be accessed in many cultural spaces. Jenkins focuses specifically on fan culture, but one can consider all aspects of media, performance, and social connectivity as being influenced by these technologies. The primary example that comes to mind when thinking about Web 2.0 technologies is the emergence of so-called social media.

Many tend to think of the term social media primarily as platforms like Facebook or Twitter (X)—because they operate through communication via mediated language and relationality. These platforms operate as "social networking" (Lonergan 2016, 26) spaces, where the primary objective is to create communicative and community networks via social bonding. Social media also refers to platforms such as YouTube, TikTok, or Instagram, where communication occurs through more performative or aesthetic types of material posted to generate subsequent engagement from a variety of participants. These spaces act as domains for user-generated content that impacts the social sphere. For example, the uploading of videos and pictures transmits information that adds to and performs as cultural and social dialogue ready to be circulated beyond the individual "wall" of the original user. These videos and pictures have the agency to change social expectations and configurations. Before the digital age, social media might refer to television, radio, or telegraphs. Even further back, social media came in the form of a book or newspaper, and before written language, it was theatre, dance, and ritual. I approach the term social media in this manner as a way of explaining how any communicative medium—often aesthetic or technological—helps transfer information or explain the world. Jenkins (2006) also describes a medium as "a technology that enables communication" and a "set of 'protocols' or social and cultural practices that have grown up around a technology" (14, quotations in original). When adding the word social to media—an assemblage of multiple mediums—I refer to a technology that uses human communication (language, writing, sound, embodiment, etc.) to exchange information in a reciprocal manner. Hence, theatre as well as other forms of interactive modes of performance like online message boards, interactive narratives, and even collaborative video games are all forms of social media.

Like the above theorists, a central concern of networks covered here is conceptualizing a network as a "proliferating multiplicity that at once enables and challenges our very capacity to think" (3). Even more so, our capacity to act and to act in ways that have political capacity. When I say politics, I mean a capacity to make change through direct actions of the individual in relation to any form of social media. When it comes to the politics of discourse, the participatory condition reconfigures the agency of the individual as one in the many who "appears before others as an equal" (Barney et al., xiii) to better understand how to give voice to the voiceless. This democratic action evokes a "re-distribution of the sensible" through the process of *dissensus* described by Jacques Rancière (2009, 62). In dissensus, a new model of participatory politics deconstructs the consensus establishment where only certain voices are given the ability to be heard. This would equate to the previous modes of unidirectional media reception and hierarchical media production practices Debord warned about. Instead, networked collaboration opens up a multitude of possible structures for emancipatory politics led by participant spectators through this new form of interconnected relationality.

The participatory condition, therefore, "confirms the possibility of equal participation by all actors (artists, spectators, curators, etc.) in the aesthetic regime" (Barney et al. 2016, xv). This participatory condition has emerged due to the interactive nature of 21st century participatory culture. The social networks that develop out of digital culture create a condition that "is disposed to participatory citizenship and collective action" (Lavender 2016, 17). Through this condition is one can break cultural norms to develop new understandings of "voices and values" (17) to re-determine and re-shape the spaces that make up our collective selves. These selves are shaped by informational flows modeled after the network of networks embedded in Web 2.0. It is through the logic of the network, an interstitial space/time configuration, that a form of collective intelligence emerges among the many different voices connected and communicating through networked collaboration.

As Web 2.0 became ubiquitous, some began to warn of over-valorizing the impacts of networked collaboration (Carpentier 2011; Jenkins et al 2016). The participatory condition prompted by interactive systems might be a negative type of hyper-individualism when approached through consumerist and capitalist means. On the other hand, it might also lead to radical forms of participatory politics and communal discourse when used in certain ethical aesthetic constraints. In the closing sentences of *Convergence Culture*, Jenkins (2006) argues that the participatory condition develops through the convergence of old and new media and necessitates a "need to be attentive to the ethical dimensions by which we are generating knowledge, producing culture, and engaging in politics together" (294). My impression is that he hoped to wish into existence a new understanding of civic engagement in the era of the participatory web. One that appeared possible during that cultural milieu of the early 2000s and seems completely lost today. Technogenesis via Web 2.0 offers the capacity to urge forth an ethico-political position when networked collaboration is applied in ethically considered ways. This occurs when the individuals involved harness participation as a form of community building via eithico-political relationality. Communities emerge through a mutual understanding of individualized agency activated as part of a larger whole. This whole always exists but is often hindered by the ideology of difference, acting as a classifying agent that comes from the humanist model of social dynamics. By thinking through a mediatized construction of sociality that includes networked collaboration and subsequently the participatory condition, difference can be thought of as a trait that can unify rather than divide. The earliest platforms and domains of Web 2.0 encouraged this new form of sociality.

Technologies of Networked Collaboration: Social Media and Participatory Tools in Online Networks of Web 2.0

Digital communication can exist as a form of participatory action when found in the many different interactive web platforms and spaces that form the backbone of Web 2.0. Described as the second iteration of the internet developed in the late 1990s, Web 2.0 promotes an ethos of user participation baked into the operational qualities of a variety of domains, software, and online platforms. Web 2.0 was based on a new digital sharing system where open API (application programming interface) protocols allowed programmers to piggyback off already existing software or pools of data to develop new work in a participatory manner (Anderson 2012, 25). With the advent of new dynamic HTML protocols, web pages became increasingly interactive and adaptable through users' direct input without the necessity of understanding coding language. Dynamic HTML helped change the relationship between the everyday user and the web page from passive viewer to interactive participant (25). With these changes, a new social paradigm emerged, allowing non-tech savvy interactors the ability to actively shape and reshape virtual worlds and develop new domains of collaboration and sociality online. This, in turn, began to reshape human perception into a technogenetically-conditioned mediatized format via networked collaboration.

The term Web 2.0 was first introduced by information architect Darcy DiNucci at the end of the last millennium and became a more common name for all user-centered internet applications around 2004. She described her vision of the next evolution of the internet as a platform founded on collaboration and participation between all users, with little top-down approaches to power dynamics. This ran counter to the earliest understandings of the internet as simply a broadcast medium and an extension of the prevailing logics of media that informed much of the twentieth century. DiNucci (1999) contrasted this new network platform to the earlier seedling internet in this manner: "The Web will be understood not as screenfuls of text and graphics but as a transport mechanism, the ether through which nteractivity happens" (32). Unlike Web 1.0, which primarily consisted of static websites or digital bulletin boards, Web 2.0 introduced interactive platforms that encouraged user-generated content and collaboration. This shift paved the way for the emergence of social media sites, platforms that transformed the way individuals and communities connect and share information online. While the earlier model was user-centered, it was still primarily sender and receiver based. The term Web 2.0 gained popularity in the early 2000s as new web-based applications and services emerged, such as social media platforms, blogs, and wikis that allowed clear reciprocal action between users in real time. What was different about these new services was how they enabled their users to create and share their own content, interact with other users, and collaborate across domains in a more egalitarian model. Web 2.0 had a profound impact on society and culture by fueling the rise of relational forms of mediated communication and collaboration. This in turn changed the ways most people learn, work, and consume entertainment in the twenty-first century.

The network of platforms, software, and domains that make up Web 2.0 is vast. A search query utilizing Google's Bard Chatbot asking for examples includes:

- Social media platforms: Facebook, Twitter, Instagram, LinkedIn, YouTube
- Online collaboration tools: Google Docs, Microsoft Office 365, Slack, GitHub
- User-generated content platforms: Wikipedia, WordPress, Medium
- E-commerce platforms: Amazon, eBay, Shopify
- Learning management systems (LMSs): Canvas, Blackboard, Moodle

(Bard Chatbot, October 20, 2023)

These examples have become the everyday backbone of nearly all aspects of mediatized life. The basis of each platform and system is interactive functionality that allows a broad-based set of collaborative tools, that puts the user in the position to drive content creation and communicate directly in a non-hierarchical manner. The platforms themselves are tools for collaboration in the widest sense of the word.

The new and growing interactive capacity of Web 2.0's foundation allowed users to become more creative in the ways that they could add their own generative content, as well as allowing them to collaborate in a variety of ways. The relationship of Web 2.0 to a person's perception of the world operates in the following manner:

The spectator point of view is so internalized that from that same perspective, individuals are able to observe their own lives, their own experiences ... Individuals think of themselves as being watched by an audience, using the set of tools and criteria of judgment they get when they were only the audience for their own narratives and communication practices.

(Napoli 2014, 189)

The impacts of this new way of thinking about how one participates in the operations of Web 2.0 bleed out into general society and can be seen in a variety of social spheres. There are various economic and political impacts directly attributable to Web 2.0. For example, social media platforms have enabled businesses to reach new customers and markets, and user-generated content platforms have enabled individuals to create and monetize their own content. You can see these last two merging in the rise of influencer culture, where social media personas join with major corporations to become personalized nodes for direct-to-consumer advertising. Political impact includes the way social media platforms have been used to organize and mobilize political movements such as the Occupy movement, the Arab Spring, and the Parkland massacre protests in the US. These movements can be seen as the positive aspects of online collaboration, but there are also negative political consequences. Social media channels have been used to spread disinformation and propaganda, as seen in multiple election cycles of the 2010s, specifically the 2016 run up to the American Presidential election and the referendum vote related to the UK's decision to leave the European Union. Social media have also been used by government entities in Southeast Asia to suppress political uprisings. These examples show how the early adoption of Web 2.0 tools began with an almost utopian idea of participation that then led to a form of manipulation as use became more sophisticated and embedded within everyday culture.

The overall social impact of Web 2.0 can be measured in multiple ways depending on the purpose of your survey, but it is undeniable that there has been a profound impact on social relationships across the digital sphere. Social media platforms have created a much more seamless way for people to stay in touch with friends and family, while creating new opportunities for people to connect with others who share their interests. In a utopian sense, the connectivity offered new opportunities to learn more about those who have different interests, offering new potentials for social relations. The Web 2.0 social media sphere is a theatrical space that allows its users to transcend the role of observer into that of actor, director, and playwright (Lonergan 2016). As a theatrical space, it empowered individuals to share their voices, connect with others beyond their immediate social sphere, and participate in civic discourse in a manner that shapes publics. The rise in social media has also been significant in the rapid rise of social movements, where the platforms operate as digital systems for activism and organization around politics of all natures. The social media sphere quickly became the performance space where participation becomes the norm, displacing unidirectional and passive forms of media consumption dominant throughout the twentieth century.

While I could spend much of this section detailing the history and evolution of Web 2.0, I will instead focus on the rise and evolution of social media platforms in the first decade of the 21st century due to the ethos behind many of their initial operations. Each platform at first offered a model for democratic world building and communication, becoming a form of experiential performance space where people had power over larger power centers of media. I focus primarily on the rise of these social platforms prior to them becoming publicly traded companies or being acquired by major tech conglomerates in order to focus on the philosophical underpinnings behind the ideas these platforms offered prior to the full commercialization of their logics. These platforms emerged in the late 1990s and early 2000s and were first characterized by their dynamic and interactive functionalities that separated them from the previous communication platforms found in Web 1.0.

One of the first sites for online networked communication was Six Degrees, founded by Andrew Weinrich in 1997. Like the party game six degrees of separation, the platform allowed an expansion of one's friend network beyond their immediate groups. The platform was based on a desire to connect people based on their real-world relationships. Users created profiles and added people they knew to profile pages, which then could be extended to the friends of these people, creating a wider network who they could then communicate with by sending direct messages. The platform did not gain widespread adoption but set a precedent for the giants like Myspace and Facebook in the coming years. Developed shortly after Six Degrees, Blogger was launched in 1999 and served as one of the first platforms to focus on giving users the ability to generate their own content for consumption by others on the platform. Blogger was eventually acquired by Google and used to establish the democratization of content creation like what is found on YouTube, another later Google acquisition. Blogger gave individual users the tools to create and share their own short form editorial style narratives we now know as blogs. These blogs expanded the role of online publishing of material into a more democratic manner by sidestepping the authority of major news organizations. The rise of blogs marked a clear shift in the way users accumulated and disseminated public knowledge. This new form of "citizen journalism" upended the conventional paradigms of news distribution. In previous modalities, only major companies had the capacity to develop and distribute content, but now individual users have gained the power to drive media narratives. The success of Blogger promoted a workable model for user-generated content and would lead the way toward microblogging on Twitter (X) and news site aggregators like Huffington Post, Medium, and Feedly.

In 2002, Jonathan Abrams developed the site Friendster, which allowed the development of profiles and groups similar to Six Degrees while allowing the posting of comments and photos within the friend network. This platform was wildly successful—at least by 2002 standards—with a wide appeal to younger users of the internet and became one of the earliest social networking sites to gain mass adoption, garnering three million active users in its first year. The site ultimately grew too quickly, and investors did not focus enough on user experience, leading to its ultimate demise in the face of another upstart with a similar mission: Facebook (Rosen 2018). Taking the cue from Six Degrees and Friendster, the mid-2000s saw the rise of what we know as the behemoths of social media, specifically Facebook and YouTube. Myspace should be considered in this grouping for its cultural impact during this time as well, but its star was eclipsed by the end of the decade. Founded by Tom Anderson and Chris DeWolfe in 2003, Myspace was specifically geared toward users in their 20s and 30s, with a tailored focus on musicians and artists, allowing customizable profiles with an emphasis on personal user favorites from across the music and film industries. The platform also integrated the blog format and served as a way to self-promote a user's work to help them build fan engagement. Myspace was my own first personal foray into the social media space in late 2006 as a way of building a following for the theatre and film projects I was involved with in Los Angeles. Pretty much everyone I knew was on Myspace by 2007, and it seemed imperative to be there in a culture predicated on networking and an expansion of your friend group to succeed artistically. Myspace's popularity peaked shortly after many of us became active users and was then largely replaced by the combination of YouTube and Facebook.

Facebook initially focused on connecting Harvard University students in the 2004–2005 academic year before expanding to other universities and then eventually becoming the public platform we know today. It had a similar user interface to Myspace but gained much more appeal due to its emphasis on building "friend" connections beyond artistic promotion. There was an early focus on privacy as well that allowed the curating of networks built out of similar likes and dislikes, not to be confused with the promotion tool of "likes" that was introduced in 2009. Facebook's continuous expansion and acquisition of other competing companies allowed it to achieve a near global domination with active users within a decade. In the platform's first year, it quickly gained one million daily active users; by 2008, when it became a cultural icon, it was at 100 million; by 2012, when the company went public, there were nearly a billion; and in 2024, there are a little over three billion (Dixon 2023; Richter 2021).

Rising alongside Facebook was YouTube, first created by Chad Hurley, Steve Chen, and Jawed Karin in 2005. The platform had one of the largest impacts on the way people consume and share videos online globally. Following the pattern of networked collaboration, users could upload, view, and share videos, providing a platform for creativity, entertainment, and information dissemination. A hallmark of the early YouTube ethos was that it served as a platform for the development and delivery of amateur video content that could challenge the dominance of legacy media. An element that would set it apart was the social aspect of the video content. Capitalizing on the participatory nature of social media sites like Myspace and discussion boards like Reddit, YouTube allowed both creators and consumers to comment on the videos which then brought the work into a larger social sphere. Like the office water cooler chats one might have the day following an especially engaging television episode on network television, YouTube allowed for instantaneous discussion. If the video posted went "viral," it allowed the comments to encapsulate the opinions of possibly millions of viewers, which democratized the ability to take part in the conversation across a much wider demographic. The variety of content streaming eventually led to a ranking system where videos could be listed as "Top Favorited; Most Discussed; or Most Responded" (Anderson 2012, 205). Like Facebook, YouTube also pushed the concept of content sharing into the culture, breaking down the domination of the ways legacy media controlled the networks of content dissemination. YouTube's popularity soared quickly, leading to Google's acquisition of YouTube in 2006. With the acquisition, it soon developed into a mega platform that is now the second most visited website globally, generating over two billion active daily users world-wide.

By the end of the decade, the logics of Web 2.0 had quickly taken over the cultural understanding of how communication works in a world under digitalization. New platforms would continue to launch that either built off previous models or introduced new ways of further engaging users through content creation, sharing, responding, and interacting through online collaboration. Two platforms that became staples for how people communicated in short format content include Twitter (now X) and Instagram. As Instagram serves as a platform that resides more clearly within the later phase of Web 2.0 with the entrance of both gamification and algorithmically controlled content, I will not cover it in any detail here. TikTok, emerging in the US in 2018, is also not covered, as it would fit in this paradigm as well (Boffone 2022). Twitter (X) is included as it operates as a bridge between the technic of networked collaboration and those in the following chapters based on its primary delivery system.

Twitter (X) was founded in 2006 and led by Jack Dorsey with the goal of merging "Internet Relay Chat (IRC), instant messaging (IM), and mobile/cell phone (SMS)" (Anderson 2012, 233). All three were "texting" formats utilized by different mediums. By combining the features of these three protocols into one platform, Twitter could capture an entire generational demographic of digital communication that would allow for direct dissemination of information using the microblogging format away from the home computer. A defining trait of Twitter was how it focused on user experience via smartphone. Launching around the same time as the first iPhone, Twitter would work to capture the energy behind the new wave of mobile internet. This helped lead to a rise in lifestyle blogging that was on the move and gave birth to the # (hashtag) tool. As ideas, content, and stories circulated across the Twittersphere, the # became the currency for virality and user connection, operating as a spotlight in a sea of information made opaque with the flood of new voices. The earliest format of Twitter offered a space for amplifying one's voice and promised an avenue to have that voice travel beyond the confines of friend groups or circles of likeness. Growth on Twitter was slower than Facebook and YouTube, taking until 2010 to gain 40 million users (Iqbal 2023), but this allowed a user base that would eventually gain a reputation for operating as a more "serious" space of information sharing. While Twitter was at first largely used to communicate thoughts and experiences about one's daily life in a form of lifestyle blogging, it quickly became a place where citizen journalism would flourish and become the go to platform for social causes and political speech. By the turn of the new decade, it became one of the most influential platforms where voices were democratized to support or denounce a variety of political causes. The platform was influential in the 2009 Iranian uprising, the variety of protests in Egypt, Tunisia, and Libya in 2010–2011 named the Arab Spring, and the worldwide protests against the 99% with the Occupy movement in 2011 and 2012 (Jenkins, Ford and Green 2013). While Twitter never gained the same level of mass audience appeal as that of Facebook, YouTube, or Instagram, it evolved into a place for both serious journalism and the echo chamber of political misinformation. On both sides of the spectrum, however, there was an emphasis on creating a system where people had more power over the information disseminated and consumed than the previous systems dominated by unidirectional delivery of information by corporate entities. That is, until the power of the algorithm became the norm, reengineering the potential of direct participation and choice.

Participation: A Re-Assembling of Spectatorial Agency

If digitally constructed social media platforms are one of the primary technological domains influencing the shape of mediatized subjectivity and subsequently a new networked and participatory social reality, then it might be useful to ask how participation can be used as a social good within performance and spectatorship. This might first happen by asking how one can harness the logics of Web 2.0. There are many opinions concerning exactly what participation is and how it works in and as performance (Harvie 2013; Hadley 2017; White 2013). In the Introduction to James Frieze's (2016) edited collection on participatory performance, he explains that "the crux of participatory performance lies not in the object of our attention, what might be normally called 'the content', but in the ways that our attention is managed, the ways in which our engagement is co-opted *with and as* content" (23, italics in original). He is clearly considering how one's attention, engagement, or interaction with and in the event helps co-create the event itself, which then becomes the content. This rightfully explains how content can no longer be separated from the entirety of the event in which the spectator takes part and contributes to. This follows the logic of networked collaboration and Web 2.0, where content creation is predicated on the actions and agency of the user/spectator. However, the totality of the circuit between dissemination, reception, and response in the digital sphere creates the eventness of the content in relation to all those in the networked space. Participatory performance is therefore a form of social media that relies on live human interaction and direct participation to model new forms of social configurations.

Gareth White (2013) explains, "Audience participatory performance has among its building blocks—its media—the agency of the participant, and their point of view within the work" (26). These building blocks are part of nearly all forms of staged performance and not just audience participatory modes, due to the inherent necessity of an audience being present in some capacity. All audiences take part in the feedback loop between the actor/spectacle and event/spectator. If they did not, they would simply be in the presence of action onstage (3). This means that audience participation indicates something more active and directed toward the audience, who contributes through their collaboration with(in) the event. This collaboration has the capacity to make change for both their own meaning-making and the meaning-making of the other participants around them. Like the other architectures offered in this project, I highlight participation as a form of experience through the subject positions *with* and *in*. White specifies that audience participation requires the inclusion of the "audience, or an audience member, in the action of a performance" (5). Here, the action largely deals with either the way the narrative unfolds or within a form of (post)dramatic (non)action within the event. There is a nuance here that must be highlighted. Participation with(in) a performance object/event suggests "a mode of involvement on the part of individual spectators" that includes "being *of* the world and *in* the world" (Lavender 2016, 26). While this beingness is not overtly political, it denotes an enhanced form of politics via aesthetic closeness, meaning-making, and communicative interaction with the capacity to bring about change. A participant's mode of engagement and agency is dependent on their subject position as either a part of the performance event or as a spectator outside the performance narrative. In both positions, the spectator is crucial, but in the case of participation, the spectator is necessary for any action to ensue. In the exchange between the product in flux, the participant has the agency to make a material impact on the event's totality.

Societal intersubjectivity developed under a mediatized condition often leads to a relational position where spectators operate in one of two ways: Either a sense of communality arises, or extreme personalization and individuality take over. Often, the neoliberal conditions that underpin technological development put these two positions in tension with each other, where the power of community is often diminished in favor of the individual. Thinking of the consequences of globalization, Baz Kershaw (1992) refers to this occurrence of fragmenting community cohesion as the "paradox of cultural expansionism" (59) that was already beginning prior to the impacts of digitalization. The paradox would strengthen as the networked condition became embedded. The fragmenting also partially explains the emphasis on individual affect discussed in relation to immersion. Considering experiential performance's increasingly precarious place as a commodity and the growing emphasis on media that rely on the experience of the indivdual, better understanding how participation allows spectators tangible agency becomes helpful. Critiquing the relationship between experiential forms of performance and commodification in culture, Adam Alston (2016) explains how participation might come in many forms, again conflating different modes of exchange under the term participation. He states,

> These modes might include multi-sensory stimulation; an encouragement to seek out something or someone; the playing of a role (either clearly specified or ambiguous); dialogue; interaction; and the performance of tasks. I suggest that these modes of participation are all geared towards the arousal of an audience members' experiential faculties. (245)

While the last part of his statement follows the argument in this project, there seems to be a wide understanding of participation as an envelope that contains multiple types of experiential exchange. Instead, a narrower understanding of participation as something based on the relational action between an individual participant and the collective media experience is useful. In this experience, agency is measured through material impacts found in direct choice making and/or communicative collaboration in the form of dialogue with(in) the total event structure. Alston's collective writings have been beneficial in critiquing the overdetermination of experiential performance as an "emancipatory" avenue for the spectator by placing it very specifically within a culture of commodification, where the individual and their unique agency prop up the reason for making experiential work. This argument helps to better understand how individual experience fits within a neo-liberal framework and limits spectator agency within a larger cultural milieu. Even so, audience participation can have many positive outcomes. For instance, as an experience, participation can allow spectators the potential to expand their horizons of thought and feeling beyond themselves, to think once again of themselves as part of a larger community; one aligned with the logic of networked collaboration. It is through this networked spatial configuration that much of the heralding of participant agency is found. While framing the potential of participation with(in) a civic framework, Amanda Breed and Tim Prentki (2018) explain:

> Agency is closely related to notions of civic engagement and refers to the capacity of participants in performance projects to be or to become social actors individually or, more likely, as a consequence of forming themselves into a collective. By agency we understand the ability to initiate action or to act upon the actions of others who are influencing or seeking to influence the ways in which we behave as citizens .... The possibility of agency is greatly enhanced when communities are able to experience and observe the consequences of their agency. (5–6)

Following this line of thinking, agency is then best understood as a crucial component within the participatory framework of civic engagement when the effects of action are seen by the individual and subsequently the collective. It is a tangible agency that has measurable impact for all involved.

A mediatized spectator operating with(in) the *Architecture of Participation* can develop ethical and communal exchange through the specific aesthetics of social performance, utilizing tangible agency to make materially visible change. These changes can take on a political dimension, allowing the performance to then operate as a conduit for social change. When the spectator has the ability to materially participate in the dramaturgy of the event, it addresses what some have discussed as a lack of political efficacy (Rancière 2007; Bishop 2012) in performance, promoting a return to an audience that engages with the performance as a direct collaborator rather than simply passive receiver of political messaging. In this context, participatory performance operates through a form of postdramatic aesthetics that forms when the spectators are "require[d] to become active co-writers of the (performance) text" (Jurs-Munby 2006, 6). In this manner, participatory performance unifies the spectator with the performance community through their networked collaboration. It can also address a mediatized spectator as part of a diverse, networked, and entangled community, allowing a form of discourse that promotes tangible agency via direct participation in the event and/or the dramaturgy. Tangible agency is reciprocal and continuous, in the same manner that dialogic communication works as a process versus an outcome. Continuous transfer incites social and political efficacy through its discourse as part of the participatory condition in contemporary society.

When thinking about theatrical participation as a form of experience that allows the individual to connect in a collective manner and to engage in agential action, it models forms of audiencing where spectators interact with both the event and fellow spectators to perform as interlocutors in social politics and civics. This is the opposite of a form of political performance where the audience simply absorbs a message from the mouth of the author. As was briefly mentioned in an earlier chapter, the rise in electrification helped to establish the binary between spectator and spectacle as one of witness and product and not an intersubjective and relational act, creating a gap between a long running continuum of spectatorship based on dialogue and civic participation. Spectators on one side of this gap operate as the sole receivers of information versus producers of dialogue. Their feedback primarily exists via psychic energy transfer from spectator to stage or vice versa. A performance event in this model internalizes the dialogic action between event and spectator. Instead of speaking with the audience, characters within the dramatic container take on the role of communication to simply transfer a message to the spectators instead of allowing them to communicate or even understand they have the agency to communicate with the total event. In this model, participation in a material and communicative manner is no longer necessary because the event operates as a medium of unidirectional transfer to the spectator, limiting their agency.

The Architecture of Participation offers a remove from the dramatic model described above. It can also be considered as one that has re-emerged out of both a technogenetic paradigm of mediatization and what Hans Thies Lehmann's (2006) calls the postdramatic paradigm. Lehman argues that postdramatic theatre is no longer "subordinated to the primacy of the text" (21) and thus operates in a different manner than the dramatic or humanist model. While not the only postdramatic aesthetic, participation offers the mediatized spectator and ability to engage in ideological discourse in the moment of participant exchange. Following Lehman, and considering the rise in participatory forms over the last few decades, the dramatic frame for the theatrical event might simply have been a gap in the long-running history of mediated communication. When an emphasis on the primacy of communication between characters (intra-scenic) recedes in favor of participant discourse, spectators regain agency to communicate with(in) the event from their position outside the dramatic narrative (extra-scenic) but still within the full event structure (128). In this mode, "theatre is emphasized as a situation, not as a fiction" (128), and communication, both between collections of spectators and this collection and the total event, takes on new importance.

Lehmann argues that theatre, "whose main principle" has always been participation, has had a "real cause for concern" because of an "emerging transition to an *interaction* of distant partners by means of technology" (167). Lehmann originally published this argument in 1999, right before the mass implementation of Web 2.0. I argue that because of these technological means—which are now more mature than at the time of Lehmann's writing—media utilizing participation gains the capacity to re-engage with audiences in a manner that allows for meaningful communication both inside and beyond the event. The participatory model engages with the spectator in more active and reciprocal ways of communication with the potential for social change. In the following second hinge of this chapter, I move on to Instances of participation in various modalities of experiential media. Each is offered as a way of reading the experience of participation through the technic of networked collaboration, where the relational capacity of various choices causes a politics of change to arise.

3.1   The Democratic Spectator: Choice Making and Political Exchange in Participatory Media

In the preceding first half of this chapter, a framework was established through which one might begin to think about and analyze participatory experiences. The focus is on the relationship between what emerged as both a networked and participatory culture, predicated by the rise of a set of technologies and media logics forming the technic of *networked collaboration*. Through this technic, mediatized spectators are read as those who seek to materially participate with(in) their performance situations, unlocking a form of tangible agency where their actions have the capacity to make an impact on either the event itself or to reshape the perception of the event in a way that both the participant and their collective peers can see.

The Instances in the following pages come from a variety of experiential media. The first focuses on interactive narratives found in digital formats to show how deliberative choice making allows for a flexible and open form of exploration. The interactive narrative game, *Detroit: Become Human* (2018), sets the stage for how an individual makes complex choices that cause divergences in narrative flow which changes both the story and overall experience. I then expand to include theatrical examples that capitalize on the incorporation of collective audiences, whose material participation creates a form of active discourse concerning the politics of choice and collaboration. The first focuses on how collective choices made by participating audiences help to guide the narrative forward in the interactive theatre production *The Twenty-Sided Tavern*. This *Dungeons and Dragons* style story is augmented through performer improvisation and online device polling that allows the audience to shape the story as it unfolds before them. An interview with the makers of this event and the developers of the polling software used shows how one might design participant experiences. While the makers like to refer to the event as a game, its theatrical setting and its mode of exchange allow me to think of it as a participatory event. The group collaboration guides a collective experience based on carefully crafted choices. After briefly describing the experiential format of the production, I include a transcript of interviews with the project's makers. They discuss how agency, choice making, and collective audience building shape the development of the production which transferred to an Off-Broadway venue in spring of 2024 with official sponsorship from the toy company Hasbro. The final instance is included as a way of connecting to a brief historical time of revolutions that can easily be said to have occurred or at least supported through the logics of networked collaboration. Through an analysis of discursive participatory politics in the mixed-media experience *Occupy Your Mind* (2011/2012) by The Civilians, I document how the experience of participation with(in) media can be better understood as a direct response to the excitement toward democratic potential found and explored in platforms of Web 2.0 that reached a point of ubiquity in the politically turbulent summer and fall of 2011. I focus on the connection between participation and the social as a mediatized construct for questioning and activating political resistance. This final instance coincides with a critical time of political unrest in the world where online participation helped to fuel and support collective forms of protest. This is a period that also marks a shift in cultural production, where the utopian promise of participation and the logics of networked collaboration seemed to give ground to the commercialistic capacities of gamification covered in the following chapter.

Instance: Remaining Human?—Choice Making and Tangible Agency in Digital Narratives

I have been called forth to assist in a hostage negotiation. My name is Connor, and I am a RK8000 series cybernetic organism built by Cyberlife, the leading company developing lifelike androids here in 2038. My model has been developed specifically to help the Detroit City Police Department investigate the occurrence of deviant android behavior. My programming dictates my choices. Upon entering the Philips' domicile, I am confronted with a woman begging for her daughter to be saved. I show no emotion and decide to talk to Captain Allen. He tasks me with assessing the cause of an android servant going rogue and taking his ward as a hostage. When not responding quickly enough to the text prompt, I am immediately pushed on toward my task. Trying to reengage shows my inaction has limited the choices I can make, being unable to access the previous choices. This is the first indication I have that my choices are going to matter. They will matter for both Conner and me, the participant. I begin searching for clues to what led to the events to come. While moving through the room, I see various icons pointing me toward possible clues. The first is a dead body on the ground, which I scan for information. Through my special forensic abilities, I am able to reconstruct a historical event sequence showing me who the person was and how they died. It is the hostage's father. He was shot by the deviant, but why? Upon learning who he was, a percentile score measuring my potential for success appears and increases to 64 percent. Making the "correct choice" seems to matter. I next choose to pick up a nearby tablet computer and learn that the deceased father has just purchased a newer model android replacement. I move to the adjacent bedroom, where I learn more about the hostage and the deviant by listening to a recording on a set of headphones. After leaving the room, I find another body, this time a police officer who has been shot, and reconstruct how this occurred. Doing so leads me to the dropped officer's gun which I pick up and go out on the patio to confront the deviant, whose name is Daniel, and his hostage, Emma.

As the cut scene progresses, I am immediately shot in the shoulder but am not phased. Remember, I'm an android, and not really human. I am then offered a set of choices bracketed under the main task of **"Saving the Hostage at All Costs"**. The next cut scene shows Daniel surrounded by snipers on the roof and a helicopter flying above. I tell Daniel my name, and he immediately asks if I am armed. My choices are **A. Truth or B. Lie**. I choose the truth and subsequently throw the gun away to show my non-hostile intentions, and my success score increases. I slowly walk toward Daniel, who is on the ledge of the roof with a gun to Emma's head. By next choosing to work in a sympathetic versus aggressive manner, I hope to gain his trust. I try to reason with him, being as truthful as possible while not pushing him to do anything rash. I explain that I understand why he is doing what he is doing but that it is not Emma's fault. He tells me he still blames her for not loving him. I explain the emotions he is feeling are just errors in the software, rationalizing his deviant behavior as a mere glitch in the system. He tells me to command the helicopters away because the noise is too much for him. I choose to do so and then reassure him that if he lets Emma go, nothing will happen to him. This entire time I have been moving toward him slowly, and he tells me not to come any closer or he will jump. My choice is to either **A. Give Up or B. Come Closer**. I choose option B. Daniel then yells at me, "I've spent my life taking orders, now it's my time to decide," and immediately begins to fall backwards with Emma in his arms. I am prompted to mash the X button rapidly while I run in slow motion toward the falling deviant, and in a last-ditch effort, grab Emma's arm while heaving myself off the building with Daniel. I manage to use my momentum to push Emma up onto the roof but fall to my own demise with Daniel as a result. While falling, I close my eyes calmly, and a Mission Successful message flashes on the screen.

"This is not just a story it is our future" (Quantic Dream). These are the first words I heard upon first loading up the game. Calling *Detroit: Become Human* a game is a bit misleading, as it is structurally a digitally delivered interactive choose-your-own-adventure story, often described as an interactive movie. It is, however, packaged as an interactive game for the PS4 console system. While there are game mechanics involved, the primary mode of exchange is based on the way one makes choices throughout the narrative and the consequences of those choices on the narrative versus making choices toward achieving a clear objective. These choices become crucial, however, as they have a significant impact on the overall story delivered to the participant. The choices drive a set of interconnected story arcs intended to help the participant gain a sense of empathy and understanding of the technological beings who have become conscious by exploring acts of free will, hence the name Become Human. The game asks the participant what it means to be human in a highly mediatized world and challenges assumptions about androids as simple machines. A key component of the experience as a form of political activation is that it does not offer easy answers to the complex themes introduced. Instead, the participant learns to think about the issues based on their choices and how each different choice leads to different possible solutions. A major theme relates to how one manages individual and collective agency. There is a political dimension to the choice making architecture as it offers the participant a better understanding of how true democratic discourse and action work. Every choice and every individual character arc is important to the collective whole of the history within the game.

A key dynamic to the game is how it uses the interconnected story arcs to build empathy in the participant based on the way they navigate their choices. By relying on this capacity to connect with the participant, final outcomes and opinions on the experience are varied, which becomes a net positive. The variety of possibilities allows for continued discourse on the way one navigates the choices in online forums, which has pushed users to replay the game over and over again to see the impact of their choices. The official subreddit for the game has over one-hundred thousand members (r/detroitbecomehuman). The game's studio describes taking part in the narrative in this manner:

PLAY YOUR PART IN A GRIPPING NARRATIVE – Enter a world where moral dilemmas and difficult decisions can turn android slaves into world-changing revolutionaries. Discover what it means to be human from the perspective of an outsider—and see the world through the eyes of a machine.

THEIR LIVES, YOUR CHOICES – Shape an ambitious branching narrative, where your decisions not only determine the fate of the three main characters, but that of the entire city of Detroit. How you control Kara, Connor, and Markus can mean life or death—and if one of them pays the ultimate price, the story still continues…

COUNTLESS PATHS, COUNTLESS ENDINGS – Every decision you make, no matter how minute, affects the outcome of the story. No playthrough will be exactly the same: Replay again and again to discover a totally different conclusion. (Quantic Dream)

Depending on the source one relies, there are between 40 and 80 primary endings available. In reality, there are far more than that based on the minute permutations that can cause incremental changes to each. In an interview with Jade King (2018), the lead writer estimated there are over 1,000 different combinations of endings for the third act alone. The sheer number of endings contributes to the possibility of agency for the participant (Chen, Dowling and Goetz 2023). To just give a brief example of how this might occur, let me return to my previous playthrough of the chapter "The Hostage" described at the start of this section.

There are six possible major endings to this first chapter (Lavoy 2018). In each, the participant controlling Conner through the narrative is presented with diverse ways of possibly saving Emma.[1]

- Connor Failed to Reach Deviant in Time
- Connor Leapt for Emma and Fell
- Snipers Shot the Deviant
- Connor Died Protecting Emma
- Deviant Shot Connor
- Connor Shot Deviant

The outcome of each has an impact on the narrative, as it triggers a software instability in Connor which will become a major plot point later in the story. Remember, to get each of these possible outcomes requires a different combination of choices made throughout this sub-section of the full chapter. Below is a sample of the decision tree flow chart of my first playthrough (Figure 3.1). You can see there are multiple un-opened branches. For example, in my playthrough, I did not interact with an injured police officer during the negotiation. That officer is off to the side during the approach, and Connor can interact with him, offering an option to help treat a bleeding wound. If you choose to do so it opens up a cross-chapter dialogue farther along in the narrative. While that choice does not impact the ending of "The Hostage," it does impact the overall story told and experience for the participant (Figure 3.1). Every choice is consequential and creates an almost endless number of possible permutations within the narrative. Each of the choices can be thought of as a nodal situation (Domsch 2013).

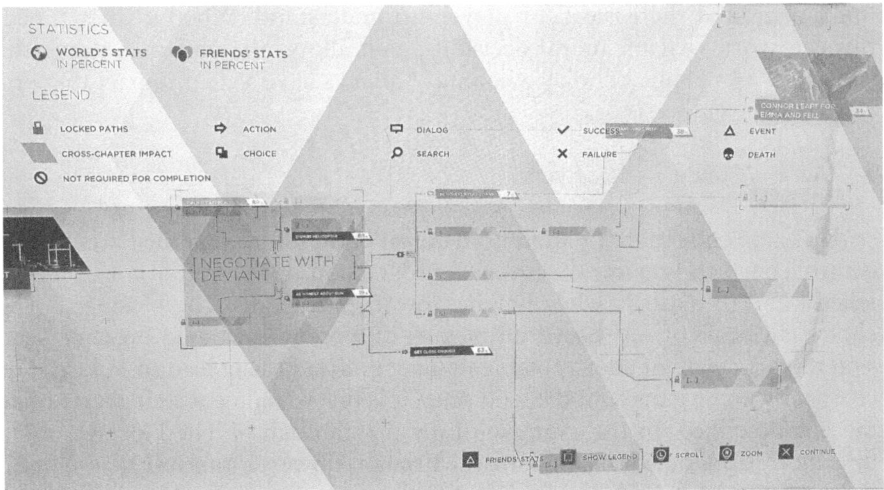

Figure 3.1 Personal Player Flowchart for "The Hostage" in *Detroit: Become Human* by Quantic Dream.

When researching participant responses to the narrative, there were various responses to how one navigates these choices. The number of Reddits, online forums, and other online posting sites extend the narrative through the logic of networked collaboration. The participant can also connect to thei friend network on the PS4 to see what decisions they made, offering a broader commentary on participant agency and the social capacity of the game. In the discussion boards, the response to an individual playthrough becomes multiplied through commentary and questioning about the myriad potentials one might encounter. These discussions are often geared toward understanding the best choices which are highly subjective. But that is the point with participation, it allows the mediatized spectator the ability to steer away from the one true understanding or meaning behind a performance. In the online space of discussion boards, however, a collective is formed due to the multiplicity of options made available by the makers of the narrative.

One of the more useful discussions I found was on the official Reddit DetroitBecomeHuman thread. User "throwawaydeviant9" laid out a systematic approach to all the ways the game offers choice-based architectures for participant experience. They state: "I've found that *D:BH* uses a variety of ways to present choices to the player, and variety of ways to demonstrate the effects of those same choices" (2018). They explain how each choice opens up a new branch. Each choice should then be considered a node within the system architecture of the narrative. The way they categorize these branch options includes:

- "One Explicit Choice"

The user explains that "this is the most straightforward example of presenting a branch to the player. It is very flexible, easy for the developer to implement, and the easiest for players to understand. When it's not timed, this can work well for moral decisions, as it allows the player to weigh the situation." An explicit choice will almost always have a clear impact that one can see later on.

- "Action/Timed Reflex Events"

These are choices that require some form of input through the participant's controller which impacts the physical progression of the avatar, for example, when I had to mash X while lunging toward Daniel in "The Hostage." The choice is less so of one based on picking an option but based on your decision to participate in an acceptable manner and can easily lead to failure. The characters in the game can die, and often it is due to failing a Reflex event like the one described. In the example of my playthrough of The Hostage, it led directly to Connor sacrificing himself. Connor, however, can resurrect himself by transferring his memory to a new body, unlike other characters in the story.

- "One Problem, Many Solutions, Few Outcomes"

The user defines this choice as "the idea here is that instead of branching a scene by presenting an explicit choice, you create high amounts of branching *within* a particular scene, which will converge to a set amount of outcomes." The entire negotiation scene of "The Hostage" is of this sort. The problem presented is **Save the Hostage at All Costs**. There are multiple options offered, starting with either telling the truth or lying about the gun. Do you keep it or do you throw it away? Throwing it away immediately eliminates two out of the six endings listed above: Deviant Shot Connor or Connor Shot Deviant. Each subsequent choice eliminates a potential ending/solution to the problem.

- "One Path with Many Fail States"

This form of choice making is often used to build tension because no matter what you choose, the outcome is negative. One of the most interesting things about *Detroit: Become Human* is how it is predicated on pushing the participant to seek out many different navigations through the story by returning to the chapter and trying new options after completing the entire storyline. The limitation of the One Path Many Fail States form is it can dissuade the participant from making choices that deviate from the path toward potential success, therefore severing branches before they have an opportunity to grow.

- "The Second Chance"

This form of nodal option often might be paired with the previous. It allows the participant to correct a previous mistake without starting all over.

- "The Culmination of Previous Choices"

This category is one of the most difficult to parse, as it requires understanding how the entire network is built and which nodes in the network can impact the others. I'd argue that *Detroit: Become Human* is partly successful due to the use of this choice system. In the playthrough above, I missed a crucial choice option that would only present itself had I been keenly observant from the very beginning of the scene. It has a significant impact on Connor's character arc. His story is largely guided through an instability in his software. The more unstable it becomes, the higher the possibility he has of being presented the option of choosing to become deviant himself. Because my choices as Connor through the entire journey were largely predicated on the fact that he was a machine designed to help the police control and investigate deviancy, versus an android with altruistic impulses, I kept him on a path away from "enlightenment." At the beginning of "The Hostage," a fish was lying on the floor in front of a smashed aquarium. I originally missed this. Had Connor seen the fish, he would have been given the option to pick it up and put it back in the aquarium. This first interaction establishes the root of Connor's instability. In terms of narrative possibility, it signals to the participant that there is a possibility for Connor to be deviant by showing empathy very early on in the narrative. The experience of gaining empathy is intended throughout the major story arcs of all the playable characters.

So far, I have only discussed the first chapter which operates as both an introduction to Connor and a tutorial to the system architecture and branching format. The next three chapters introduce you to the other two primary protagonists of the narrative, Kara and Markus. Markus begins as a caretaker of a well-off elderly man in a wheelchair named Carl. Interestingly, it is Carl who has genuinely cared for his android by teaching him/it to read and to make art. Markus is shown what love is as Carl has obviously taken on the role of father figure. This introduction to Markus establishes even further the question of what makes a human based on the choices one makes. In a later chapter, Markus is nearly destroyed while trying to protect his owner, and his sacrifice starts him down the path of becoming deviant. That path becomes the core of the narrative, as Markus represents the awakening of all future androids to becoming human. Markus is a symbol for the power of political choice. His journey leads him to become the savior of the android masses, who can eventually become awakened to their power and autonomy, which leads to a revolution against their human overlords. Though you might become empathetically connected to Markus, his story is one of measured decisions as they inform a future imbued with political agency.

Kara, on the other hand, is an android servant and nanny tasked with housekeeping and childcare of Alice, a small girl being raised by a meth addict named Todd. If Alice manages to survive her first major altercation with Todd, her story serves as a parallel to Markus's but is more focused on the agency one has in interpersonal relationships and love versus political agency. Unlike Connor, who can be rebirthed anytime he is killed, Kara can die in her second scene, severing her entire story branch. If she lives, she goes on to become a deviant herself and a mother figure to Alice. Kara and Alice's journey through the game focuses on seeking refuge from the abuse of humans like Todd. As their narrative unfolds, the choices presented are guided by how close or distant Alice is to Kara. The more Alice begins to open up to Kara and subsequently the more Kara becomes a mother to Alice, the participant learns how it is possible for machines to love. Later in their branch, the participant learns that Alice is an android too, but by this point, they have likely transferred the bond of care that Kara has for Alice onto themselves. This relationship building is crucial as it impacts how the participant makes the choices offered to Markus and Connor. How they choose determines the ending where an android uprising leads to either a collective union in trust between humans and the machines or utter destruction of one or the other.

Overall, the goal of creating the interconnected storylines is to "invoke empathy on the players' side for the characters they play" (Shubert 2021).

Participant empathy and agency become intertwined, allowing for an ideal form of politics to emerge. Returning to Jagoda (2016) from the first half of the chapter, the experiential nature of *Detroit*'s narrative represents a form of political and ethical participation through the logics of its nodal structure. By engaging its inherent network aesthetics, "it helps us experience ethics not as a restrictive code but as a collaborative process that opens up various, if often unsuccessful, ways of being and becoming together" (216). Through the agency of the participant and their critical inclusion in shaping the narrative, a potential future is explored. *Detroit: Become Human* is more than a game, and even more than just an interactive narrative; it is an exercise in democratic agency both informed by and structured around networked collaboration.

Instance: Rolling Dice and Deciding Fates – Designing Participation for Audience Agency

There is a raucous energy in the room. The audience members are dressed in their favorite fantasy and sci-fi gear and are ready for the adventure to begin. They have just been introduced to the interactive nature of the experience to come when asked to participate in a drinking game with Sarah, the game master, and the other adventurers. A lucky participant walks up on the stage. Will the participant drink ambrosia (apple juice) or poison (pickle juice)? The audience gets to decide. This drinking game operates as a short tutorial on how to use the Gamiotics polling software the audience members have connected to on their mobile devices. Not only are they swiping and pressing buttons to help decide who drinks what and how much, but they are also advised to use the devices to take pictures and videos throughout the show and to absolutely share those memories on their social media channels. The lights go dark except for the faint glow of the screens, and the adventure begins. This will be unlike most experiences of gathering in an audience space to see a theatrical production. From the start, the audience is shown how their agency to participate will mean something.

Nearly a century has passed since the ritual of the crystal cauldron was last performed. The magic done that day created the world in which our story takes place. Where order and chaos clash in equal combat, though the good often prevails. Legend spread about a climactic battle, and the spooky ostentatious platypus that was defeated there. We are called by the adventurers of the past to follow in their footsteps. For the time of the ritual is upon us again and the story must be written. But what dangers will arise, and how will they be faced? Who will join us on our journey, and what will their legacy be? What pathways will open and how will they lead to the heart of this ever-changing labyrinth? It's time to find out at The Twenty-Sided Tavern!!

The Twenty-Sided Tavern is a part choose-your-own-adventure, part improv production, and part community building event that uses digital device-delivered audience polling to shape a narrative about mages, tricksters, and warriors who work with the audience to solve the mystery of the crystal cauldron. [2] As they say in the intro, *"The fate of the story, lies squarely in your hands."*

The following is an edited conversation with the creators of *The Twenty-Sided Tavern* David Carpenter, Sarah Davis Reynolds, and David Andrew Greener Laws.

Will Lewis (WL):   What is your perspective on audience agency?

Sarah Davis Reynolds (SDR):   You can't say audience agency without also saying authenticity, that's sort of the way that I've always looked at what audience agency is. We joke a lot that the illusion of choice is still a choice, but that can't always be true with live things, especially. In order to give the audience agency, you do have to sometimes trust them and not railroad something. This is especially true when working on a *D&D* sort of thing. That can be scary, to really leave the choice up to the audience and not decide how things are going to go. But in order to create that authentic experience, we do that, we throw that at the audience and are like, "you get to tell us what to do and we have to go along with it." It's a lot about the authenticity and then honoring the choice that they've made. You can't ever say, "you made this choice, but we're actually going to do this instead." Because then that makes it inauthentic.

David Carpenter (DC):   In terms of telling a narrative, where instead of the author saying sit back, let me tell you a story, you are put into the driver's seat of the telling of that story. Your choices and your actions are deciding the narrative as it's going, as opposed to you being told the narrative. And so, when you look at stage entertainment, the playwright isn't necessarily telling you what to think. But they are telling you what to see. You're supposed to walk away with your own kind of impressions, thoughts, ideas, and reactions to what the playwright presents. You can, however, create a whole experience where it's up to the audience. For example, I've gone and seen brilliant work that was marred for me, like plays or musicals that were marred for me because I had a bad day. So, it made me ask, what if we do something where you don't separate out the emotion of the moment, or you don't separate out what's happening, but you're actually using that in the experience as an audience member and that is determining the outcome of what it was. Then, of course, not punishing them for that. So, all these things combined are where this concept of agency really came in. Rather than telling the audience a story, we ask, why don't you participate with me in the creation of the story, and let's see what happens?

David Andrew Greener Laws (DAGL):   Audience agency is very rare in the theatrical world, and very important to all of us that have come to *The Twenty-Sided Tavern*. Whether we come at it from a gameplay perspective or a Shakespeare perspective, we are doing this show for these people who are here, and we are having this experience with them. We might as well give them something to do as well, so they're not just sitting in statically and simply watching. I think that one way to break it down is in traditional theatre, the audience has the agency to react, and we're really focused on giving the audience agency to act. So instead of just watching whatever is happening and then reacting, whether it's oh, there was a joke, I'm gonna laugh, or instead of injustice, I'm gonna stand up but not actually affect what is happening. We're interested in giving agency that does actually act and change the way that whatever is happening in the next five minutes is going to happen.

DC: I'm thinking the stage, the storytelling, the craft of storytelling on stage. With this interactive idea, it is just ripe for what you bring with you tonight. You are going to define the experience. It's taking the power and sharing it with your audience. That's it as an artist and a creator in a way that most playwrights I know would hate. But that's what I love about it. That is who we are; we're asking them to the party, you know, in a totally different way, and giving them a voice in a totally different way.

SDR: Something that we really focus on with *The Twenty-Sided Tavern* is also presenting levels of agency so that the audience can interact in the way that they are most comfortable. For example, a traditional theatre audience doesn't want to scream NPC names at us, but they probably can become comfortable with sitting in their chair and voting on their phone. Because it's not as loud or, you know, as spotlight driven. So, we really have a whole range [of ways to participate] by saying, "Here's a menu. What would you like from this?" Basically, "How much agency and interaction do you want to have with us?"

WL: *That's really great! So next question. What is the overall intended experience for the audience member? Or the audience as a collective group?*

DAGL: For me, the goal is for each individual audience member to recognize how essential they were to the story we were telling. The recognition that the story without your presence would have been completely different. Either because you did vote in the winning way, or didn't vote in the winning way, or said a name or didn't say a name. Even nonparticipation changes the energy in the room. And it wouldn't be the story that we told that day, without you, the individual there.

SDR: When it all boils down to it, we're throwing a party, and we want you at the party with us, so we're all just having a fun time. You know, that's something that we use. And we've had to sort of talk to our marketing team about, it's not just a show that they're seeing, we are inviting them to have a fun time and party. Our actors are out in the house during pre-show and getting to know everyone. And so, if we were just putting adjectives on the experience, it is community-based. You are coming home to a community who you feel comfortable with and just having a great time.

DAGL: We want them to go on an adventure. We track the hero's journey for the actors on stage and for the audience. From the time they hear about the show to them leaving the experience and interacting with it after the fact, we want them to have changed and have grown and have become heroes through doing this experience together. Hopefully experiencing something new or unique or special or exciting.

WL: *Who would you say is the ideal or imagined audience for The Twenty-Sided Tavern?*

DAGL: There are two kinds of ideal audience. There is the audience that is ideal for us as performers, which is a bunch of *D&D* nerds, because they love it, and they're going to be loud, and they're going to be ridiculous. They're going to play the game in a way that no one else is going to play it. And then there's the other kind of audience that is people who've never seen *D&D* before, people who've never experienced interactive theatre before, people who don't know what this is, and don't know what to expect. Because, you know, I come from a big Shakespeare background, and nothing thrills me more than someone seeing a Shakespeare play that I put on and then going, "I didn't think I was going to understand a word, and I loved it." I think we have a collective desire to convert people to our way of theatre. The ideal is both a mix of people who've been doing this their whole life and people who've never experienced it, so they can all have a joint understanding of what's so exciting about this.

SDR: An ideal audience is one who is open-minded and receptive to anything.

DC: So, we took *The Twenty-Sided Tavern* to Pittsburgh last year and it was our first developmental production. We partnered with Pittsburgh CLO and we said just put out Facebook and digital advertising, and only go after *D&D* fans and gamer fans. Right before we started, we started getting questions on our website, saying, "Hey, I see this is a *D&D* style show. Can we dress up?" And we're like, "yeah, please do." And so, these early performances, people started dressing up in costume. Most of it was like *D&D* Ren Fair type stuff. But as word of mouth started getting out, our sales started to go up, and up and up every performance, we were selling more than we did before. People continued to come dressing up, and by closing weekend, the audience looked like the floor of Comicon. And this isn't an exaggeration. It really did. Because you had Ren Fair, you had *D&D*, *Lord of the Rings*, *Magic the Gathering*, *Star Trek*, and *Star Wars*. You literally had representatives from multiple fandoms, right across sci-fi and fantasy, right there. What we realized is that we had created a safe space for nerds within the live entertainment spectrum in a way that no one had ever done before. This is the world that they want to live in. This is the world that they want to inhabit. And there's an enormous audience of them out there.

That's my goal, of course. That's what I want to be doing with my life. I want to celebrate who I was as a kid and pull this into mainstream entertainment. We're using this interactive software to say, you don't have to sit and watch a play. Don't get me wrong, I love watching plays. But there's more, there's another way to be entertained. One that mirrors the experience you have as a mobile user from birth.

WL: *Can you expand on the digital tools you used to craft the story and the ways you use digital logics to develop the participatory elements of the show?*

DAGL: There are a number of digital tools in Gamiotics' arsenal, from moments of trivia to riddle solving via text entry, from the sheer tenacity of the multiclick to the group finesse of an over/under. For *The Twenty-Sided Tavern*, we have assigned each skill on a hero's character sheet to a specific election: Multiclicks for athletics, over/unders for acrobatics, etc. This way, the challenge we're facing in-game is emulated by the task we are completing in the space. The performers can communicate with the audience, trying to influence certain decisions, steering them toward or away from different tools based on the skill set they know their character has. But ultimately, it comes down to which path the group wants to take (Figure 3.2).

Then there are other moments of heightened tension or intense decision making where very specific tools are used only for such an occasion. There is little more satisfying than the result of a full, digital tug-of-war during an immensely significant decision. By suiting the action to the circumstance, we develop another layer of the audience's ownership over their decisions and, ultimately, their outcomes. When things go well, they are their own best

Figure 3.2 Performance still of audience choosing Trickster character from *The Twenty-Sided Tavern*. Photo by Bronwen Sharp.

champions. And when things go poorly, they are bolstered to new successes in the future, must rally to perform better as a unit, and will have to face the consequences of their own performance in the meantime.

WL: *In what ways does the audience participation shape the creation and the outcome of the work? I'm thinking both from the developmental perspective and possibly on an individual night perspective.*

DAGL: I would say somewhere in the middle of those is how frequently the audience composition changes the show as an experience. We go to a different city or a different country, and different bits land differently; jokes will or won't hit, references will or won't be understood. Sometimes it's very evident when we do the show in New York versus doing the show in Edinburgh. Sometimes it requires a different comedic styling. But sometimes it'll be going from Pittsburgh to Chicago. There will be a decision that will overwhelmingly one way. And we really don't know why, there are just so many factors. Whether it's the way a word is being said or something's being lit or how a design has been made. And who knows what's influencing these decisions? Maybe it's a cultural thing of like the Left versus the Right. Who knows? And so, in that way, the audience is really important. We're always writing and rewriting the script. The script is never settled. We're always looking at, hey, this decision could be more equal. Could this choice be more compelling? What can we do to adjust it?

SDR: We really consider it a living work. It's never locked or frozen. And the beauty of that is that we can take that feedback and that input that we are getting from the audience and adjust things.

WL: *I'm wondering then, how much agency did they have to make the story better or worse. In terms of the overall experience.*

DAGL: I think Sarah and I agree that whatever experience they have, it should be satisfying and earned. Regardless of what happens, it's not unfair what they get. Sarah and I are huge proponents of thinking that nothing is locked behind chance; there is always an opportunity to get a "perfect" ending. People are coming back because it is so different every time. Even if you get a perfect experience, maybe you want to come back and see the worst experience, or maybe you want to come back and see something else.

DC: I think that as an author, I don't ever want to punish my audience for showing up and having a bad day, right? I want that at the very least, they should leave with some modicum of hope, even if it's a bad ending, right? It's not their fault. They showed up to play a game. They played it badly. And you want them to come back? So, part of it is this relationship that you build with your audience. My choice is, I want to have a relationship [where] I want you to come back, [where] I want to give you hope, even if you have a bad day.

SDR: There's so many narrative elements and character arcs that you can't really ascribe perfect or failure to. So, people are really coming back to that as well. They might have had a "perfect game," but they want to see different characters, they want to see different bits. And all that changes the experience.

WL: *Have you encountered any instances where the audience has broken the game?*

DAGL: The joy is that anything can happen. And you can use it, whether it's coming from the audience or coming from elsewhere. Like the roof could cave in, and we could still use that in the adventure. The difference between something like this and something like a video game is that if you fail at *The Twenty-Sided Tavern*, you still have to continue. If you lose in a video game, you can start over, and it can be a ten-minute experience or a ten-hour experience. This is always going to be a set time. It's going to be a two-hour show, but you still have to have the capacity to fail. So, whether they're doing it on purpose or whether they are not succeeding accidentally, there has to still be a story that continues. And so that's why I think it would be difficult for them to break it.

WL: *I think that's a great ethical way to approach it as an artist because you could say, well, I'll give them the bad ending, and then maybe they'll come back the next night. But I think a lot of corporations would like to approach things like that in certain ways. Like, with Netflix's Bandersnatch.*

DC: The problem with *Bandersnatch* was that your choices didn't matter. Your choices were not very consequential. They were simply directional. They didn't really impact the narrative in a big enough way throughout the course of it that made me feel like it was important that I was there. If the choices don't have a consequence, then there's no point to it. That's the point of a choice. If you want to give a choice in a narrative and you're going to give them the agency, then there must be another side of that, that reflects the choice that got made in a serious narrative manner.

        When you look at *Twenty-Sided Tavern*, there's one choice in the show you saw that I absolutely hate. At the end of the show during the escape, when they're leaving, there's a moment where you meet two NPCs. And the audience has to choose one or the other. It's a great choice because we're giving the audience a life-and-death choice right at the end of the show, which is really good because life and death is the most important part. So, we do this life-or-death choice, and we save someone, and it has no impact for the rest of the show. You take the most important possible choice you can make in life, which is life or death, and then not follow through with it. That was one of our big lessons. We'll never make that mistake again.³ But it's also part of what I instill in the [Gamiotics] platform. You can have small choices, you can have big choices, we can create a way to build in choices, dramatic tension, but you can never have an inconsequential choice. It's not fair to the audience; you're breaking the contract, if you're giving them agency, you have to actually give them agency. You can never set up a choice that feels like you're telling the audience which way to go. You have to actually say, "It's up to you, what do you want to do?" and then live with what the audience chooses.

        So, from a design platform perspective on Gamiotics. We have to design choices for an audience that feels fifty/fifty. That feel evenly balanced so that they don't feel short changed, that they see how their choice actually matters. That's how we keep them engaged. We really reinforce over and over again; your choice does matter. It does matter. It does matter. It has consequences; it does matter. We do it over and over again.

WL: *Last question. How do you, as theatre makers, as game makers, as people interested in interactive forms of storytelling, see audience participation, changing the overall expectations of what we call theatre moving forward?*

SDR: This is a difficult question because I think we internally say over and over again that what we're doing is not theatre. But I also know that it is a helpful benchmark to grab on to. If we are putting entertainment on a stage, then there is inherently the comparison to theatre. There are definite shared elements of it. That being said, like we just talked about with the audience, you know, if it is an audience member who is coming in expecting a red curtain theatre experience, that's not an ideal audience for us. In the same way, you know, that person is probably not going to enjoy any sort of audience participation or interactive theatre. There is a group that what they want their theatre experience to be is just a passive, I go watch something. And so, we say a lot that what we're doing is not theatre, but I don't think that it's not theatre. I think, especially with the pandemic, but even before that, we're becoming a much more active society. And that the way people want interaction, people have shorter attention spans, we're given all sorts of choice, you know, the world lives here for us now. I think as artists and as art makers, it is our responsibility to respect that and react to that and embrace that and sort of innovate into that.

DAGL: If this isn't theatre, I do hope theatre can still take a page out of our proverbial book. I am entirely anti-fourth wall. If we could do a piece of theatre without the audience changing the show, we might as well do it at home in our living rooms. It might as well be TV at that point. If their presence is not going to affect the story, I don't see what the point of doing it live is.

WL: *Do you have anything else you feel like you'd love to add?*

SDR: At the end of the day, everything we do with *The Twenty-Sided Tavern* is for the audience. You know, really, truly every decision that we ever make. What town do we want to set this in? What currency are we using? What characters are we doing? How are we casting this? Every single decision that we make or question that we ask ends with. "What is this mean for the audience?" I don't know if traditional theatre does that in the extremes that audience participation in theatre does.

Instance: (Re)Performance as Civic Activation

2011 was a pivotal year for revolutionary participation across the world. It was also a monumental year marking a shift toward the ubiquity of the participatory web. At the peak of the political potential of the participatory condition, the sentiment of a potentially lost generation, brought about by the economic and social devastation of the Great Recession spurred an increased interest in media that explored the dynamics of the 99%. In the fall of that year, The Civilians, a theatre company devoted to civic engagement through journalistic performance, developed the project titled *Occupy Your Mind*. The part verbatim-cabaret performance, part interactive digital protest, staged and continues to restage the rhetoric and actions of the Occupy movement using material participation in the form of (re)performance via

YouTube and online blog. The company offered the project to give spectators the means to rehearse and perform ethical- and community-based dialogic action. The Civilians' project used social media to negate a "slacktivist" action of merely watching and engages a different type of agency and political efficacy through a spectator's transformation into participant and direct conduit of democratic messaging. The act of offering the audience to partake in the performance harnesses the logics of networked collaboration while encouraging direct action in the political events that seemed to engulf Western and specifically American society that year.

The Civilians' mission is to "investigate our lived experience, interrogate the stories that shape our society, and awaken new thinking through the experience of live theater." (Civilians, n.d., "About") Similar to the style of Anna Deveare Smith or the Tectonic Theatre Project, much of their work comes from a journalistic perspective and strives to tell the stories of real people and real issues. The Civilians' verbatim work is emblematic of the postdramatic due to the way the final product in performance serves as an attempt to "create an open-ended journey rather than a conclusive story with a beginning, climatic middle, and resolution" (Kozinn 2010, 192). For *Occupy Your Mind,* the company created an open forum for the political angst and unrest that came to fruition in the Occupy protests. This forum was both live and mediatized. The performances began as a set of interviews taken from members, participants, and onlookers of the Occupy camp in Zuccotti Park. From the over 150 interviews transcribed, a select number of curated monologues were performed as part of the ongoing cabaret series *Let Me Ascertain You* at Joe's Pub, part of the Public Theatre, on October 28, 2011. In the spring of 2012, the group took these transcribed monologues and made them available to the public as the next wave of disseminating the political messaging. The collection of verbatim monologues was also filmed and archived on a Tumblr page (Civilians 2012) created by the group, which allowed a larger cross-section of spectators to interact with the content. That spring, the Occupy movement was slowly vanishing from the general public's perception, partially based on its lack of media presence. Many seriously motivated activists continued to spread the movement's message through direct action events or digital channels, but much of the lay public had moved on to other topics of interest steered by the mainstream media.

To keep Occupy's message in the social consciousness of the public, The Civilians turned to the participatory web by using the medium of YouTube. Media theorist Henry Jenkins (2013) describes YouTube as *spreadable media* that "functions in relation to a range of other social networks; its content gets spread via blogs [...] Facebook and Myspace, where it gets reframed for different publics and becomes the focal point for discussions" (275). The Civilians used the platform as a way of connecting the message of the performance to the many nodes in an ever-expanding participatory media ecology. The performances could move freely about the digital sphere by participatory interaction such as commenting, sharing, posting, remixing, and most importantly, (re)performing.

A performative call to action was formed by The Civilians, whose purpose was to "capture the living history of the movement as it unfolds" (Wallenberg 2012, 28). To do this, the group asked digital participants to take up the mantle of the protest and become the next generation of political agitators and commentators by (re)performing the monologues from the Joe's Pub cabaret via the digital archive. Not only were spectators offered agency through their own (re)performances, but they were also prompted to find other Occupiers in their local vicinity and interview them for future performance and protest opportunities. This call initiated a continuous feedback loop of spectator as participant engagement. The Tumblr site gives instructions on how to conduct new interviews, including release forms and tips for the best ways to get one's subject to feel comfortable with them as an investigative journalist. On the one-year anniversary of the September 17, 2011, "start" of the Occupy movement, the Civilians also staged a (re)performance of the original monologues in coordination with over 14 other performing arts organizations. The event was titled *Occupy #17* and included performances of the transcribed material by both amateur and professional performers. These performances can be seen on the Tumblr page and heard on the group's podcast (The Civilians, n.d. "Let Me Ascertain You").

The way in which participants can engage with *Occupy Your Mind* has the potential to open-up a new vista of political agency through the emancipation of the spectator. By emancipation, I mean the spectator is freed from the bonds of information transferred from one object (a performer or the spectacle itself) to another (themselves) as a form of communicative mediation. Instead, the spectator is free once given the opportunity to enact the information themselves. Returning to Lehmann (2006), "The consciousness of being connected to others and thus being answerable and bound to them 'in the language,' in the medium of communication itself recedes in favor of communication as (an exchange of) communication" (184). This mode of communication is therefore relational instead of unidirectional and requires feedback. The invitation to participate in the online discourse by (re)performing the verbatim interviews gives the participant a new sense of connection with the real potential of the original political event. Participants enact the communicative force of the original act instead of simply absorbing that force by only listening or watching like they might do when watching a news report or reading info from a news feed. They adopt the logic of Web 2.0 tools by responding through their own material participation in the creation of a future performance event.

When a spectator is invited to enter into the moment by participating, they gain agency by "treading the borderline, by permanently switching, not between form and content, but between 'real' contiguity (connection with reality) and 'staged' construct" (Lehmann 2006, 103). The effect of the (re)performance has increased potential due to the direct engagement with its mediatized spectators. By extending the production of the performance out into the digital sphere, a plenitude of newly embodied co-authors emerges. No longer is the message relegated to that of spectacle meant for consumption, but rather it is assimilated and activated as a form of political agency for the spectator. Steve Cossen of The Civilians explains that with their form of verbatim theatre, spectators can "actually strip away our overly narrow preconceptions of how people work, how the world works, how social systems work—whatever the subject is" (Kozinn 2010, 196). In accessing a politically activated Real through the non-fictive "post-drama," the participants gain a new way of seeing, conceiving, and realizing the political implications of the work. The invitation to (re)perform the event allows political affirmation through the conduit of participant agency. A spectator turned into participant enters a political Real by reliving the emotions and thoughts of the original via (re)performance. When the spectators become the reality through their own performance, they gain direct access to tangible political agency. It is the doing, the performing, the act of being, that releases the energy and efficacy in the moment. The participating spectators have the capacity to shape a new reality through their performative actions of embodiment.

I refer to the Civilians' project as both participatory and postdramatic due to its re-positioning of the spectacle/spectator dynamic through the use of spectator material engagement combined with mediated dissemination. The use of the participatory form offers an increase in political efficacy and spectatorial agency. The purpose of direct action is to continue the dialogue concerning participatory democracy that the content discusses. The content's potential is accessed by a (re)performing of a verbatim dramaturgical text that depends less on the spectator's gaze than on the participant's action as emancipated performer and interlocutor of democracy. The spectacle guides the participant via performance through their mediatized choice making versus simply an authoritative direction. As guide, the text becomes the subject of a gaze in which the interviewer/documentarian of verbatim practice becomes emancipated. This gaze then allows multiple levels of interpretation and mimetic assimilation that can lead to a finished political affect in the corporeal (re)telling and (re)performance. It is through the (re)performing of the Civilians' verbatim spectacles, as opposed to the watching of a fictive narrative, that "such disturbing experiences could be called political in the sense that they evoke a de-naturalization of the dominant perspective at work in the conceptions of reality" (Bleeker 2011, 48). Through the real-life utterances forming the base of the Civilians' performance, spectators can engage with a political sphere not accessible through purely fictive drama meant for watching. This connection and subsequent agency increase exponentially through practices that ask the spectator to become one with the narrative through their (re)performance aesthetics. Using direct spectator interaction via the networked platform of YouTube, The Civilians' project keeps Occupy's message active through its perpetual feedback loop of live and mediated transmission. This is discourse that engages community and individuals alike. Through this postdramatic dramaturgy and remediation, the monologues performed allow for a potentially effective and affective way of engaging with the Occupy movement's message well beyond its original historical impact.

Conclusion: Networked Sociality and Participatory Politics

In the closing paragraphs of Claire Bishop's *Artificial Hells* (2012), she claims that participation in performance and art has at its core a goal of engaging with the social through the direct engagement of the spectator. She explains,

> Participatory art aims to restore and realise a communal, collective space of shared social engagement. But this is achieved in different ways: either through constructivist gestures of social impact, which refute the injustice of the world by proposing an alternative, or through a nihilist redoubling of alienation, which negates the world's injustice and illogicality on its own terms. In both instances, the work seeks to forge a collective, co-authoring, participatory social body—but one does this affirmatively (through utopian realisation), the other indirectly (through the negation of negation). (275)

The Instances detailed here follow this logic and serve as representatives of a mode of experience brought about by the technic of networked collaboration. This technic is still impacting today's audiences, but one might argue that it reached its pinnacle at the end of the first decade of the twenty-first century. The unrest and revolt that erupted during 2011 brought about the first faint glimpse in many years of a formation of potential revolutionary communities operating under the logic of networked collaboration. In light of the Arab Spring, the Global Recession, and the Occupy movements, these divergent communities began to gain agency and political power using the tools of Web 2.0. They formed through understandings of necessary interconnectedness brought about by historic events and the progress of technological impacts. These communities were made up of divergent constituencies who have rarely been able to form cohesive unity due to their differences and a lack of developing a clear yet distributed method for communication. It took catastrophic events to form the temporary coalitions of togetherness based on truly democratic idealism. The Occupy movement's factions could work through difficult discourse to create a horizontal governing structure based on the good of the whole. This structure mirrored many of the utopian ideals embedded in the networks of Web 2.0.

Spectator experiences found in this section harness the power of direct participation through live and mediatized channels to engage the public as networked members of a fluid democracy. By capitalizing on the potential of the participatory condition, these instances offer the power to rethink the relationship between the spectator and their media, opening a variety of potentialities for politics and ethics in our technologically saturated realities. As I continue to progress through this project, I will discuss the complicated nature of digital technologies as companions to mediatized sociality, based on how they often contain hidden dangers masked by the promise of experiential exchange. Just like the participatory condition developed out of Web 2.0, digitalization brings with it utopian dreams that can turn into dystopian nightmares under both democratic capitalism and autocratic governance. Considering all these possibilities as part of the network of sociality is necessary in a mediatized social structure. Using the instances in this section, I have argued that the logics of a social world mediated and mediatized through digitally-constructed social media allow for a potential to re-invigorate the political capacity of spectators engaged in participatory exchange. This participatory condition transposed its logic and function through the Architecture of Participation, and when harnessed for ethical, political, and communal purposes, allows spectators to gain tangible agency to re-shape society.

The plethora of opportunities and choices that came about with the rise of the internet and then democratized through Web 2.0 would eventually lead to a paradox. Could too much choice or agency to act lead to a paralyzing and polarizing condition in society? While the rest of this project will not attempt to answer that question directly, you will see how the logic of networked collaboration continues but also has been co-opted by major social/economic forces in a manner that limits those choices while still making it seem as though a mediatized spectator is still largely in control. Through gamification and algorithmic regulation, corporate and political forces began to put guardrails on the ideals of unlimited choice and unfettered agency. In the final moments of my interview with David Carpenter concerning *The Twenty-Sided Tavern*, I asked if he had any last thoughts, and his response offers an ideal segue into the next section. He exclaimed,

What I'm after is the gamification of daily lives. The reason being is, I'm engaged when I'm playing a game. I'm engaged when I have a task. There's a reason why Escape Rooms are so popular. Because I'm going go to entertainment that challenges me and puzzles me and I have to solve it. I want to win, and I want to do that. I think that concept of winning and losing, or the concept of competition within society is actually really much stronger than we give it credit for. And so, the idea of taking your competitive nature and making it fun and gamifying it is what I grew up with, and what I love. I love this idea of being able to do it in a live space and in the human space. Take this idea of gamification and move it into a whole other realm where it hasn't been possible for because technology hasn't been there. But now everyone's walking around with Penny's smartphone in their pocket. Penny's notebook from *Inspector Gadget*, right? This is Penny's notebook from Inspector Gadget now in our pocket, and now everyone is walking around with one harnessing that as a tool and using it within the entertainment spectrum.

With the advent of the mobile device, Web 2.0 became portable and a whole new generation of mediatized spectators were introduced to a paradigm shift in how they interact with(in) their media. This new mode of experiential exchange would now be gamified.

Notes

1 For an example video walkthrough of each see: https://www.youtube.com/watch?v=khPZKMZzZTY
2 The initial interviews were conducted shortly after the early iterations that had toured in various cities changing slightly for each venue. Once the production opened Off-Broadway with the official licensing of *Dungeons & Dragons* the quest changed to "Carriers of Chaos."
3 This choice was removed from the Off-Broadway iteration.

Bibliography

Alston, Adam. 2016. "The Promise of Experience: Immersive Theatre in the Experience Economy." In *Reframing Immersive Theatre: The Politics and Pragmatics of Participatory Performance*. Palgrave.

Anderson, Paul. 2012. *Web 2.0 and Beyond: Principles and Technologies*. CRC Press.

Bard. 2023. "List examples of Web 2.0." Google.

Barney, Darin, Gabriella Colman, Christine Ross, Jonathan Sterne, and Tamar Tembeck. 2016. "The Participatory Condition: An Introduction." vii–xxxix. In *The Participatory Condition in the Digital Age*. University of Minnesota Press.

Bishop, Claire. 2012. *Artificial Hells: Participatory Art and the Politics of Spectatorship*. Verso.

Bleeker, Maaike. 2011. *Visuality in Theatre: The Locus of Looking*. Palgrave Macmillan.

Boffone, Trevor. 2022. *Tik Tok Cultures in the United States*. Routledge.

Breed, Amanda, and Tim Prentki. 2018. "General Introduction." In *Performance and Civic Engagement*, edited by Amanda Breed and Tim Prentki. Palgrave.

Carpentier, Nico. 2011. "Contextualizing Author-Audience Convergences: 'New' Technologies' Claims to Increased Participation, Novelty and Uniqueness." *Cultural Studies* 25, no 4-5: 517–33.

Castells, Manuel. 2004. "Informationalism, Networks, and the Network Society: A Theoretical Blueprint." In *The Network Society: A Cross Cultural Perspective*, edited by Manuel Castells. Edward Elgar.

Castells, Manuel. 2010. *The Information Age: Economy, Society, and Culture Volume 1: Rise of the Network Society*, 2nd ed. Wiley-Blackwell.

Chen, Lei, David Dowling, and Christopher Goetz. 2023. "At the Nexus of Ludology and Narratology: Advances in Reality-Based Story-Driven Games." *F1000Research* 12, no. 45.

The Civilians. n.d. "About." http://www.thecivilians.org/about/mission.html

The Civilians. n.d. "Let Me Ascertain You." Soundcloud. https://soundcloud.com/thecivilians/sets/the-civilians-let-me-ascertain.

The Civilians. 2012. "Occupy Your Mind." Tumblr.Com. 2012. http://occupythecivilians.tumblr.com

Cook, Amy. 2008. "Interplay: The Method and Potential of a Cognitive Scientific Approach to Theatre." *Theatre Journal* 59, no 4: 579–94.

Couldry, Nick, and Andres Hepp. 2017. *Mediatized Construction of Reality*. Polity.

Debord, Guy. 1995[1967]. *The Society of the Spectacle.* Translated by Donald Nicholson-Smith. Zone Books.

DiNucci, Darcy. 1999. "Fragmented Future." *Print Magazine* 53, no 4: 221–22.

Dixon, Stacy Jo. 2023. "Facebook: Quarterly of Monthly Active Users Worldwide 2008–2023." Statista.com

Domsch, Sebastian. 2013. *Volume 4 Storyplaying: Agency and Narrative in Video Games.* De Gruyter.

Frieze, James. 2016. *Reframing Immersive Theatre: The Politics and Pragmatics of Participatory Performance.* Palgrave.

Galloway Alexander R, and Eugene Thacker. 2007. *The Exploit: A Theory of Networks.* University of Minnesota Press.

Hadley, Bree. 2017. *Theatre, Social Media, and Mean Making.* Palgrave.

Harvie, Jen. 2013. *Fairplay: Art Performance and Neoliberalism.* Palgrave.

Iqbal, Mansoor. 2023. "Twitter Revenue and Usage." *Businessofapps.com.* https://www.businessofapps.com/data/twitter-statistics/

Jagoda, Patrick. 2016. *Network Aesthetics.* University of Chicago Press.

Jenkins, Henry. 2006. *Convergence Culture: Where Old and New Media Collide.* New York University Press.

Jenkins, Henry, Mizoko Ito, and danah boyd. 2016. *Participatory Culture in a Networked Era.* Polity.

Jenkins, Henry, Sam Ford, and Joshua Green. 2013. *Spreadable Media: Creating Value and Meaning in a Networked Culture.* New York University Press.

Jurs-Munby, Karen. 2006. "Introduction." In *Postdramatic Theatre* by Hans-Theis Lehmann, translated by Karen Jurs-Munby. Routledge.

Kershaw, Baz. 1992. *The Politics of Performance: Radical Theatre as Cultural Intervention.* Routledge.

King, Jade. 2018. "*Detroit: Become Human* Can End in More Than '1000 Different Combinations,' Says Quantic Dream." *Trustedreviews.com.* April 23. https://www.trustedreviews.com/news/detroit-become-human-can-end-1000-different-combinations-says-quantic-dream-3459139

Kozinn, Sarah. 2010. "Discovering What We Don't Know: An Interview With Steve Cosson of the Civilians." *TDR: The Drama Review* 54, no 4: 188–205.

Krotz, Friedrich. 2008. "Media Connectivity: Concepts, Conditions, Consequences." 13–32. In *Connectivity, Networks and Flows: Conceptualizing Contemporary Communications*, edited by Andres Hepp, Friedrich Krotz, Shaun Moores, and Cartsten Winter. Hampton Press.

Latour, Bruno. 2005. *Reassembling the Social: An Introduction to Actor Network Theory.* Oxford University Press.

Laurel, Brenda. 1992. *Computers as Theater.* Addison Wesley Publishing.

Lavender, Andy. 2016. *Performance in the Twenty-First Century: Theatres of Engagement.* Routledge.

Lavoy, Bill. 2018. "Get 100 Percent Completion and Unlock All Endings in the Hostage of *Detroit: Become Human.*" Shacknews.com. May 24.

Lehmann, Hans-Thies. 2006. *Postdramatic Theatre.* Translated by Karen Jurs-Munby. Routledge.

Lonergan, Patrick. 2016. *Theatre & Social Media.* Palgrave.

McCutcheon, Jade Rosina. 2013. "The Performance Mirror." 145–58. In *Embodied Consciousness: Performance Technologies*, edited by Jade Rosina McCutcheon, and Barbara Sellers-Young. Palgrave.

McCutcheon, Jade Rosina, and Barbara Sellers-Young. 2013. *Embodied Consciousness: Performance Technologies.* Palgrave.

Napoli, Antonella. 2014. "Social Media Use and Generational Identity: Issues and Consequences on Peer-to-Peer and Cross-Generational Relationships: An Empirical Study." *Participations: Journal of Audience and Reception Studies* 11, no 2: 182–206.

Quantic Dream. 2018. *Detroit: Become Human.* https://www.quanticdream.com/en/detroit-become-human

Rancière, Jacques. 2007. "The Emancipated Spectator." *Artforum International* 45, no 7: 270–81.

Rancière, Jacques. 2009. *Dissensus.* Translated by Gregory Elliot. Verso.

Richter, Felix. 2021. "Facebook Keeps Growing". *Statista.com.* Feb 4, 2021 https://www.statista.com/chart/10047/facebooks-monthly-active-users/

Rosen, Jake. 2018. "Antisocial Media: The Rise and Fall of Friendster." *MentalFloss.com* September 13.

Shubert, Stefan. 2021. ""Liberty for Androids!": Player Choice, Politics, and Populism in *Detroit: Become Human*" *European Journal of American Studies* 16, no 3: 1–18.

Throwawaydeviant9. 2018. "How Detroit: Become Human Creates a 'Branching' Narrative. *Reddit.com*

Wallenberg, Christopher. 2012. "To Catch the Conscience of a Nation." *American Theatre Magazine.* https://www.americantheatre.org/2012/04/01/to-catch-the-conscience-of-a-nation/

White, Gareth. 2013. *Audience Participation in Theatre: Aesthetics of the Invitation.* Palgrave.

Wittell, Andreas. 2008. "Towards a Network Sociality." 157–82. In *Connectivity, Networks and Flows: Conceptualizing Contemporary Communications*, edited by Andres Hepp, Friedrich Krotz, Shaun Moores, and Cartsten Winter. Hampton Press.

4 The Architecture of Game Play

4.0 Pervasive Connections: Smart Devices, Locative Media, and the Gamification of Reality

Imagine a time when you did not have access to anyone at any time in nearly any place. Imagine when you spent more time looking at the actual world around you versus the virtualized representations of that world contained within a glass and metal rectangle that neatly fits in the palm of your hand. Imagine a time when the virtualized maps guiding you through that device were physical and required you to trace and plan your path physically. Imagine a time when connecting with your friends also required you to traverse that map space and find an actual location to create conversation. Imagine a time when conversation required reciprocity in real time and not just delayed responses sent through texts and reactions. Imagine a time when you were not under a constant barrage of information flowing through that device and into your sense of awareness. How does imagining that make you feel? I bet for many of you reading this, that feeling is one of nostalgia and longing for a time before you were permanently tethered to all places and times simultaneously. For the rest, you probably shrink back in terror, as if imagining that possibility is like severing your sense of self in two.

This chapter unpacks the technic of *pervasive connectivity* that creates hybrid realities formed through the fusion of virtual and actual worlds. It arises through mobile technology and the imaginative properties of game-based performance structures. In previous chapters, actuality and virtuality were discussed in terms of conscious representation of felt experience and imagined possibility. In this chapter, actuality is expressed primarily through notions of physical and geographical space as well as the corporeal actions of mediatized spectators which become altered and augmented by virtual and digital domains. These newly formed hybrid actualities are explored in relation to the technogenetic properties of locative media and analyzed through theoretical implications of a mediatized subjectivity brought about through our pervasive connections to digital technologies that are always connected. This connection refers not just to networked linkages of the internet, but also a cojoined interdependence where the technology symbolically grafts itself

DOI: 10.4324/9781003398813-5

onto the perceptual apparatus of mediatized spectators. This reduces the conceptual binary between virtuality and actuality, as the two become forever entangled through the actions of mediatized spectators who constantly traverse both Reals at once while interfacing with mobile technologies which I refer to as iDevices. Hence, it is no longer productive to understand the two as a binary or set of opposites but instead as two conceptual figurations that are forever intertwined and navigated through aspects of gamification.

The focus is on the effect and affect of mobile digital technology on the understanding of selfhood, as well as the implications of this understanding for spectatorship in an era of pervasive media. The relationship between one's perceptual apparatus and these devices urges forth a dynamic where spectators enter a game-like liminal state of techno-embodied perception and operation with(in) the world. By better understanding how individuals are technogenetically altered by this dynamic, we learn to create conceptual modes of analysis that help us understand spectatorship with(in) the *Architecture of Game Play*; a framework that offers mediatized spectators the ability to navigate these liminal states as a relational process that allows ludic critical modes of exchange via gamified extensions of agency. In ludic critical exchange, mediatized spectators encounter virtual domains developed through the combination of the fictive, imaginative, and digital to gain critical awareness of non-virtual realities they encounter daily through acts of game play.

Game Play works as a mode of experience where spectators gain heightened introspection regarding the actual worlds they live with(in). It does this by linking virtuality and actuality into a hybrid space that is held together by gamification. The blurring of these realities occurs when people are pushed to seek out rewards, achievements, and objectives that normally would seem intrinsically motivated but are brought to the foreground using game mechanics embedded in iDevice software and hardware. Gamification has become a societal construct where all aspects of life—from entertainment, to government, to education, and healthcare—have been imbued with aspects of game play brought about by pervasive connectivity. Game play operates as an in-between form of experience appearing where virtuality and actuality overlap and blend as it offers potential for future action that becomes actualized through acts of experiential spectatorship. Like the experience of VR tech in Chapter 2, game play also allows spectators to perform in the realm of virtuality while located in the physical, material, and corporeal spaces and times of actuality. The difference here is the virtual spatial dimension is ever-present without the need to "transport" using augmented visual/auditory/haptic technologies. There is no need to step into the virtual; instead, mobile tech fades into the background while working as mediators between Reals and allows the virtual to symbolically engulf all actions within actuality. As an always-on and always-performing technic employed through mobile devices, pervasive connectivity evolves the way a spectator's perceptual apparatus operates into one based in a world of "liminal hybridity" (de Souza e Silva 2009).

To set an understanding of the ways the Architecture of Game Play operates, the first half of this chapter explains the technic of pervasive connectivity and the technologies spurring it forward. A brief history of these technologies is explored to offer the reader an understanding of how mobile devices evolved from simple transmission devices for disconnected communication into a network of interlinked tools augmenting human perception that has been capitalized upon through gamification in society. This technic offers a lens for analyzing the ways that spectatorship operates as a mode of exchange within the architecture. By introducing a concise history of iDevices in conjunction with theories of mobile device embodiment, I explain how mobile technologies operate and describe the disruptive impact of the technology on human perception, ways of being, and society at large. This impact helps to create a technologically informed mode of mediatized subjectivity.

The term iDevice is introduced as a way of symbolically linking the rise of pervasive connectivity to the generational cohort of spectators coming into being alongside the rise and embedding of locative media in twenty-first century society. This cohort, more often described as Generation Z, is also referred to as the iGeneration due to their affinity toward mobile device interactivity. An iDevice is more than a simple digital tool, it is a hand-held extension of a mediatized spectator's physical and cognitive being; an extension that allows connective access to the entirety of the social world. This is even more true for those born into and raised in the era of pervasive connectivity. While the structures that iDevices establish in gamified constructions of perception are accessible without the technology, this section foregrounds how the device has become an inescapably attached part of both mediatized societies and mediatized selfhood. Once these forms of technology create connections between their user and the world, disconnecting is nearly impossible without considerable negative consequences to a conception and perception of a stable sense of selfhood.[1] The interlinking nature of the technology destabilizes one's ability to consider a singular self as possible. As a spectator, this destabilized sense of self causes the user to move between the dual Reals of the virtual and the actual, enacting a form of play. Because the device has become an always-present tool and mediator of these multiple Reals, it transforms the spectator into a *player* of the in-between.

Pervasive Connectivity and the Power of the Mobile Interface

As our tools evolve, so does our sense of perception. Nicolas Bourriaud (2002) discussed the rise in a form of digital sociability that marked a fundamental shift in the ways one perceived the world at the end of the second millennium. He warned of "epistemological upheavals (concerning new perceptual structures), stemming from the appearance of technologies" (66). As discussed in the Introduction, Bourriaud was writing about a shift toward relationality (social interaction) in the art world in the late 1990s and how that shift was partially informed by late-twentieth-century technologies. The technologies that Bourriaud refers to belong to the subset mentioned in the previous two sections. I will move on to focus on changes in relational perception beginning during the mid-2000s brought about by the pervasive influence of locative and mobile communications technologies, or what I've previously referred to as iDevices (Lewis and Johnson 2017). These technologies have enacted a technogenetic process on the members of societies whose structure is dependent on pervasive connectivity.

When augmented by the pervasive connectivity of mobile technology, people gain access to place, space, and time in a manner that transcends conventional modes of watching and even participating to develop a perceptual function that is more akin to being forever suspended in bounded yet liminal play. This form of play extends the framework of immersion by becoming ubiquitous and inescapable but manageable through structures of gamification. Due to their portability and pervasiveness, mobile technologies have a unique capacity to perpetually shape and reshape the idea of an individual's centered placement in the world. The technology travels with its user, constantly shaping their understanding of a unified self and encouraging a becoming part of multiple mediatized selves existing in overlapping pseudo-realities. When the tech merges with and augments one's perceptual apparatus, their sense of selfhood incorporates a multiplicity of material and immaterial objects that exist in the parallel worlds of the virtual and the actual. These worlds become one as mobile technology both invades and surrounds the spectator's perceptual sphere, allowing a form of heightened relational subjectivity to emerge by melding itself to the symbolic shape and functionality of the technology. The connection developed through the symbiotic link to iDevices expands one's perception of the world to allow a tethered connection to multiple aspects of time, space, and action. The technic of pervasive connectivity spurred on by smartphones, smart watches, tablets, GPS trackers, and other iDevices ingests the virtual and the social, creating a paradigm that is always on and always accessible to mediatized spectators through actions of game play.

Unlike the technologies discussed in the previous chapters, devices that operate in mobile configurations belong to the subset of computational technologies called ubiquitous computing. With the invisibility and seamlessness of computational processes these technologies bring, they are also discussed as pervasive (Farman 2021, 6–12). Ubiquitous computing technologies bring the internet and the digitally connected world to the user, as opposed to the user coming to the technology. The goal of ubiquitous computing is to make technologies and the connections they bring invisible and integrated with(in) social worlds. A working definition of ubiquitous computing or ubicomp for short is as such: "… a sociocultural *and* technical thrust to integrate and/or embed computing pervasively, to have information processing thoroughly integrated with or embedded into everyday objects and activities, including those pertaining to human bodies and their bodily parts" (Ekman et al. 2016, 5). Smartphones operate as multi-purpose machines and, as such, they capture the combined power and processes of all machines connected to digitalization. They not only connect their users to digital and virtual realms, but they also link all domains of digitality into one mobile technogenetic apparatus for technical interrelatedness (Couldry and Hepp 2017, 53–56).

Personal access to the internet has become increasingly dependent on mobile devices, which fundamentally impacts the way users interact with the world via the hybrid spaces and the interactions created in these spaces.[2] Working with mobile media theorist Adrianna de Souza e Silva (2009), Jordan Firth (2015) explains that hybrid space is formed via mobile technologies out of these three elements: Social interaction, digital information, and physical space (8). This hybrid space has become the quotidian location where many acts of daily behavior occur, causing a shift in one's sense of being and perception among iDevice users into a form of continual play. These users live in dual Real's unmoored from static ideas of space, place, and time experienced before the ubiquitous and mobile computing era. Mobile device interaction changes relationships with(in) physical spaces by altering the proximal connection between the user and location (Farman 2021, 19). A mediatized spectator's perceived experience of contemporary reality is therefore based on a dynamic in which the human body is culturally inscribed by the mobile media interface. In deeply mediatized societies, an iDevice operates not only as a tool to use but as a technological apparatus figuratively grafted onto the body of its user and into the user's sense of selfhood. This grafting brings about what Farman calls the *sensory inscribed body*. Like my own conceptualization of the perceptual apparatus, this body is a substrate that gains its mark in a perpetual and pervasive state of communication and embodiment augmented by hybrid realities introduced through the screens of mobile technologies. This body is understood as one "that is not only conceived out of a sensory engagement across material and digital landscapes, but also incorporates sociocultural inscription of the body in these emerging spaces" (13). This body is inherently mediatized because it is reified in both virtual and actual Reals via its relationship to place and space accessed through iDevice. Farman argues, "we are living in a time in which realms of the realized and the realizing (or the actual and the virtual) do not signify themselves as exclusive spaces; instead, the interaction between these spaces continues to become mutually constructive" (50). Under the influence of iDevices, the ontological basis of perception and embodiment is reconfigured and requires greater attention to the phenomenological states of mediatized spectators.

A dominant cultural force enacted upon and shaping the perception of today's mediatized spectators is a mobile social interface that moves in tandem with its user and enacts a force of perpetual movement on the user. Locations and spaces are rarely fixed in today's spectator's perception of reality. This unfixity disrupts the individual's ability to unplug from a pervasive and ubiquitously mobile sense of location in the world. It also impacts the ability to focus on one form of prescribed reality seen in previous forms of media. Constant proximity to multiplicity deters the possibility of singular focus. Farman situates this cultural paradigm shift brought on by augmentation, via virtuality, by extrapolating Heidegger's *dasein* ("being" or "the to-be") out to refer to being-in-the-world. He continues,

> Mobile technologies' impact on the production of space demonstrates how the virtual is always understood as a state of being that is intertwined with a state of becoming. This "being-as-becoming" is a present-tense experience of embodied space informed by past and future potentials. Essential to this experience of virtual space is the way that the practice of materiality is informed by various modes of representation. (39, quotations in the original)

The encroachment of shifting ways of perception partially emerges through constant connection to the virtual that has become Real through the actualizing interactions and connections to iDevices. The key to understanding the influence of iDevices is "less about the devices (used) and more about the activity" (1). This statement helps to solidify the framework for how we can think through the technic of pervasive connectivity to better understand the actions and responses of spectators within experiential performance. The ways they behave within these architectures can be read through the lens of the specific technic.

By being pervasively connected to the virtual, the ontology of space is altered by interrupting how it is interacted with, interpreted, and understood. Further discussing connections between the virtual and the actual under paradigms of digitalization, Farman (2015) argues,

> Doubleness and multiplicity of experience is key to understand what makes the virtual powerful. It is not a simulation of the real, nor is it a replacement for the physical; instead, it is an augmentation of the physical by offering experiences of the non-tangible elements that are often fundamental to life in the material world. (107)

Farman's argument helps define the impact of iDevices on the phenomenological apparatus of mediatized spectators, including the generational cohort labeled the iGeneration discussed later in this section.[3] His theory is built off the notion of phenomenological embodiment (Merleau-Ponty (2013 [1948]), that operates as a personal cultural materiality formulated through "our entire experience of the world [that] is embodied and that this embodiment frames our every perception and thought" (Freshwater 2009, 19). This material phenomenological perspective coincides with the ways that Couldry and Hepp (2017) approach the conceptual and actual impacts of mediatization on the formation of the self and, likewise, the social realities that self is in relation to.

The material aspects of the technology bring about a phenomenological intersubjectivity with its users, altering their perception of the world in a manner that co-opts the technological operations of the iDevice, expanding one's perceptual apparatus to the entirety of the world. Developing my own model over the spine of Farman's, the mediatized spectator then operates in a perpetual liminal state between virtual/digital/fictive and actual/analog/real spaces brought about by the expansion of worldly perception. Negotiating these two yet multiple spaces engages the spectator in perpetual game-like actions of embodied and virtualized interactivity. A mediatized spectator therefore operates in an architecture bound by the rules established via the technology but also free to explore the performative world by using the technology.

As argued earlier, immersion allows the spectator the feeling of agency, which I describe as affective. In participation, the participant-spectator gains tangible agency through an ability to make significant changes to narrative and event, and potentially the entire social sphere via democratic choice making. The player (playing spectator) deploys critical agency in game-based performance events by matrixing both the affective and the tangible within play. Game play combines the affective and tangible registers of experience in the other two modes with the possibility of a meta-agency of critically reflexive choice geared toward outcomes with respect to a structural understanding of the game-world. In game play, the spectator's exchange function is based on becoming a critically activated member of the event experienced due to an established set of rules and objectives. Through game play, a mediatized spectator has increased potential for consequential agency and self-determination with(in) and beyond the performance event. This potential creates what Stefan Herbrechter (2013) describes as new "possibilities of interactivity, self-representation, communication and 'identity work,'" producing "new forms of subjectivity… dissociated from material forms of embodiment" (25, quotations in original). The experience of the mediatized spectator's hybrid self and hybrid body in game play operates as a fluid and shifting mediator of multiple realities.

From Communication Device to Everything Machine:
A Brief History of Locative Media

According to the Pew Research Center (2021), 85 percent of all American adults owned a smartphone in 2021. That number has almost tripled since the survey began in 2011. Since that time, the number of other portable digital devices has also increased. In 2021, 53 percent of adults owned a tablet computer.[4] When considering the Millennial and iGen age range, the number is around 95 percent. During this time, American adults have also seen their percentage of home computer ownership decline from 88 to 77 percent, indicating an increased reliance on mobile connectivity. In 2021, most people under 65 used their smartphones mainly to connect to the internet and launch apps versus using them as a voice communications tool. That year, smartphones were used for voice communications only ten percent of the time (Flynn 2023). This is a monumental change from how the first cell phones were utilized. Like most technologies covered in this project, their initial use and functionality have been remediated. Phones became cellphones, which then became mobile computing devices over a brief time. Between 2008 and 2022, the number of hours spent accessing digital media via iDevice vs computer went from 12 to 70 percent (Flynn 2023; Henderson 2017). In 2016, 77 percent of the Millennial and iGen population had broadband internet access, a decline from the 81 percent high, showing a trend toward using their iDevices as the primary access point to the internet (Pew 2017). The number had retreated to 70 percent four years later (Pew 2021). As of 2022, 59 percent of global website traffic has shifted from physical location-based connections to mobile connections (Flynn 2023). That is nearly double since 2015 and projected to be 76 percent by 2025 (Handley 2019). These numbers show how the digital ecosystem impacting users has become mobile, and through that shift from land-based to mobile connectivity, it has become ubiquitous and pervasive.

While the history of the cell phone is over three-quarters of a century old,[5] the advent of iDevices is relatively new. The story of the smartphone begins in 1993, but it was not until the later part of the 2000s that the impact of iDevice technology became integrated (Greengard 2015). The first portable consumer smartphone was the IBM Simon Personal Communicator. The device had rudimentary access to email, a calculator, a space for digitally written notes, and a calendar, on top of voice calls. It was also the first device with a touchscreen and QWERTY keyboard. The Simon was not a commercial success partially due to its price tag of $1099 (nearly $2300 in 2024 dollars) and partially to its bulky design. The product also had serious limitations due to the processing infrastructure in place at the time. Like the early introduction of internet communication technologies of the early 1980s, it was not a product for the average consumer. At the time the device first arrived, most had yet to have much exposure to internet connectivity or email, dooming the device from the get-go.

The first true wave of early iDevice technology arrived with personal digital assistants (PDAs). Introduced by Hewlett Packard and Palm in the late 1990s, these hand-held devices served as portable digital storage systems through which users accessed limited processing capabilities to manage calendars, contacts, and written notes. One of the hallmarks of early PDAs was a stylus that gave the user the tactile input found on paper systems. By replacing paper virtually, the stored material became more portable and took up less physical space. A PDA was my own first exposure to mobile device technology. I had a Casio digital black book in 1997 where I began storing all my contacts. This was replaced with a Palm Pilot, which I was fortunate to gain when found unclaimed from the work lost and found around 2001. Scholars of locative media (Firth 2015; Hjorth and Richardson 2014; Wilken and Goggin 2015a) argue that the cognitive effects of mobile digital technologies began with these simple devices. iDevices began to operate as technologies for offloading physical human memory into digital archives (Firth 2015, 56–58). The mobile devices helped accelerate the paradigm of distributed cognition, where technological tools operate as an individual's secondary and, in some cases, primary memory bank (Firth 2015; Hayles 2012). Before these systems, many relied on their individual cognitive capabilities for memory. As the technologies become more pervasive, they slowly replace human memory systems and become necessary for many aspects of daily life. This is one of the impacts of pervasive connectivity. Just think of the last time you tried to memorize someone's phone number or address or could easily map your way to a new address in your own town. It's no longer necessary because you have a digital tool to do it for you. That tool has become part of your own distributed sense of self via its memory storage functions and connection to the networked world.

Shortly after the saturation point of PDAs, the Blackberry, made by Research in Motion (RIM), became popular for businesspersons due to the feature of immediate access to email while on the go. RIM was the company behind the software system built into the IBM Simon, but they revolutionized the field and propelled mobile device interactivity through their proprietary delivery email system. This system could "push" a user's emails directly from the server to the device on a timed protocol, which cuts down on processing and storage load of servers. The "push" software protocol is an early version of digital notifications found on nearly all operating systems today. Its introduction marks a significant historical moment when the beginnings of gamification merged with mobile devices. With "push" protocols, users no longer needed to log in to an email server for emails; instead, compressed data making up those emails are sent from the server to the user on a timed system—either constant or on cycles—making them readily available and creating a unique symbiotic relationship between digital time and actual time. "Push" technology offers one way that users become symbolically interconnected with the digital sphere at all times and all places and offers one of the first tangible moments where the operations of gamification begin to enter the picture. The use of "push" also marks one of the monumental changes in the way people connect to and with(in) virtual systems. For the first time, digital information came to the user without the user initiating the exchange. This system of delivery without the need to order operates as the first dose in the addiction of digital rewards that gamification is built upon. This addiction would only grow with the development of new models of connectivity developed as the iDevice evolved.

In 2002, RIM added cellular phone connectivity to certain models, and in 2003, the Blackberry became the first fully integrated smartphone with mass consumer appeal. Part of its popularity was the interface with its stable physical keyboard and trackball, but the idevice could also connect to the internet via new 2G wireless connections that allowed reasonable speed and more robust access for the time period. In 2004, most cellphone makers followed RIM's smartphone model and began to roll out devices with the same connectivity, but they began to push for the adoption of Wi-Fi protocols as alternatives to slower cellular networks. Through Wi-Fi, users could gain a more stable and speedy connection. This shift also allowed the providers of access to the broadband spectrum the ability to rethink the monetary model driving the growth of these devices. As devices became more "smart," they also became more data intensive, and the companies selling access began to monetize data consumption versus simple cellular connectivity (McNish and Silcoff 2015). These advances opened a space for smartphones to become a growing necessity for societies ingrained with(in) digitalization. While the evolution of the technology was rapid, the devices were still primarily used for voice and text messaging capabilities until the end of the decade.

The mass shift from mobile phones used primarily as telecommunications devices to portable pocket computers largely began with the introduction of the Apple iPhone in 2007. The way the iPhone integrated seamless interface design and cross-platform connectivity changed the world forever, opening the media landscape to new possibilities (Greengard 2015, 28). Because of the iPhone's design and functionality, along with its infrastructure for apps, the multiple connective properties of the internet became immediately accessible and integrated into a single device. iDevices began to operate as convergence machines, seamlessly integrating multiple forms of media and communication via operations of remediation (Wilken and Goggin 2015a). The iPhone was a commercial success because it was not marketed as a phone at all, but rather, as a mobile device for accessing the internet and all the applications that could run in tandem with digital app ecosystems. The launch of the iPhone also marked a moment when the full capabilities of the technology became marketable for mass consumption. Cellular phone technology made the logical but momentous leap from mobile communications technology to mobile computing technology, thus becoming "smart." With the delivery of the iPhone 3G in 2008, Apple delivered increased location awareness with assisted GPS technology for services such as directional maps. Maps embedded in a mobile device marked another important shift in how these technologies worked with their users.

GPS technologies had been part of the digital landscape for much of the first decade of the twenty-first century. Products like the Garmin GPS map tool for automobiles were incredibly popular, but they, like car phones of the late 1980s and early 1990s, were situationally specific to the space in which they were deployed. While mobile, following the flow of traffic within the moving vehicle, they stayed with the vehicle when the user parked. This aspect of the device meant they had limited use in terms of locative capacity and became extensions of the vehicle and not of the human users themselves. When the iPhone launched, it included a mapping app that was powered by Google Maps software but was limited in terms of its overall functionality. It did not include the useful turn-by-turn instructions that we are used to today. The early Android versions had this feature, but at the time, Android-based phones had not gained the market share they have today. The real change came with the 3G model, which included GPS allowing the Map app to track the user, instantiating a virtual referent to their physical travel through actual space. As time progressed, the mapping feature became the second nature way of exploring the unknown and known alike, allowing the hybrid merging discussed in the previous section. Farman (2021) describes the phenomenological impact of GPS maps embedded in the smart phone as "technological proprioception," where the user's sense of their body in physical space begins to extend into the virtual space of the digital map. That extension eventually becomes a symbiotic tethering as iDevices become part of the everyday infrastructure later in the mid-2010s.

Beyond the introduction of locative awareness within the iDevice, the 2008 iPhone 3G model also included the first access to the App store. This digital marketplace operates as a repository for new functionalities in an interconnected ecosystem based on mobile connectivity. What might seem a simple addition from our 2024 eyes, the App store marked the final shift of the iDevice from a portable communications device into a powerful mobile computing system. Capitalizing on the already existing framework of the iTunes store that helped make the iPod such a success, the App store transferred all the functionality of home computers and gaming systems over to the smartphone. It became possible to untether one's need to connect via a land-based digital device for all forms of productivity and leisure activities. Paired with the locative capacities of the smartphone, all aspects of digital life could now be on the go and always connected to the mobile user. The successive iterations increasingly emphasized storage size, enhanced sensor capacity, better cameras for both photo and video, new processors, security features, and eventually artificial intelligence (AI). At the time of this writing, the newest iPhone, the iPhone 15 Pro, has integrated advanced augmented reality (AR) capabilities with Lidar photometry. The A17 pro chip built into this model also boasts a sixteen-core neural engine that runs its bespoke AI processing capabilities. This tech is intricately linked to the Internet of Things (IoT) and processes of datafication discussed in the following chapter.

Apple followed up the success of the iPhone with the 2010 release of the iPad,[6] a portable tablet with internet connectivity that replicates the functionality of the iPhone minus an emphasis on voice calling features. The iPad, and other tablets from competitors, created a bridge between personal computer users and smartphone users. The iPad and its clones also ate into the market share of personal e-readers due to their internet connectivity. As of 2016, more people began to read digitally on multipurpose tablets and smartphones than on dedicated e-readers (Perrin 2016). With the mass adoption of tablets, smartphones, and wearable tech such as the Apple Watch and FitBit, all of which connect to the internet, the iDevice ecosystem became fully integrative and interconnected (Firth 2015, 39–42). When connected to the vast digital network living in the virtual cloud, iDevices become central nodes and create a tangible link between their users and the entire world. By entire world, I mean the vast interconnected hybrid domain made up of both the virtual and the actual. As Sam Greengard (2015) states, "The web of connectivity and interconnectivity is an order of magnitude more powerful than anything that has come before it. The technology is nothing short of revolutionary" (29). The impact these technologies have on the perceptual apparatus of their users fundamentally shifts what time, place, and space mean for mediatized spectators.

Extended Proprioception: Smartphones and iGen Ways of Being and Perceiving

Along this journey, I have tracked the relationship between technologies and human perception to better understand changes in contemporary spectatorship. Doing this has helped build an argument for a model homed in on a contemporary condition based on mediatized inter-subjectivity. Until this section, I have primarily targeted technologies established in a period relative to the social conditioning of those born before the advent of the figurative digital divide. I'd like to take a moment to return to my initial anecdotal remarks in the Introduction regarding perceptual changes based on generational cohorts. That small child, the one who perceives the size of actual things as expandable using only two pinched fingers because that is how she would do it on her iDevice, her changed perspective and relationship to the actual world has the capacity to completely upend all paradigms of spectatorship. She represents mediatized spectators that might require multiplicity and performance driven by constant modes of experiential exchange brought forth by a technologically augmented inter- and transdisciplinarity.

Because the iDevice is a late addition to the technogenetic field of digitalization, I would like to highlight its influence on a younger generation of spectators who are just now reaching adulthood in technologically advanced social systems. In this section, I narrow the focus of mediatized spectators down to a specific generational cohort labeled the iGeneration by psychologists Larry Rosen (2010) and Jean M. Twenge (2017, 2023) and more commonly discussed as Gen Z. Rosen's work on the learning capacity of late-millennial and early-iGen students further defined the term. I emphasize the effect of mobile communications technologies on this cohort because they are the first to reach adulthood having been exposed to the technology during their formative adolescent years. Twenge (2023) explains, "Although people continue to change throughout their lives, our fundamental views of the world are often shaped during adolescence and young adulthood, making the younger generations a crystal ball for what is to come" (3). The iGen's sense of selfhood is rooted in culture(s) subsumed by the internet delivered via iDevice. These devices act as primary technological markers for this generation. Although often framed by common traits based on age ranges, generations are best understood as temporal-cultural signifiers with multiple determinates and blurry boundaries. No individual member of a generational cohort will behave exactly the same as others within the grouping, but when taken in aggregate it is possible to define some generalized characteristics that will work for the whole. One way to approach these characteristics is by studying their technology and media usage. Different researchers argue about where the cohort begins, some believe as early as 1991 but 1995 is often a common start date. Utilizing the 1995 date, Twenge (2017) explains that the iGen members "grew up with cellphones, had an Instagram page before they started high school, and don't remember a time before the Internet" (2). This statement is crucial when considering the technogenetic capacity of iDevices. Following the statistics in the previous section, the iGen is also the first cohort where a majority of access to the internet comes via mobile devices. The child mentioned at the start of this project belongs on the cusp between iGen and the next cohort, Gen Alpha, also referred to by Twenge (2023) as the Polars.

The technological capacity and connectivity of the iGen mark an interstitial space where the process of technogenesis increasingly becomes co-directional. Previous mediatizing technologies of the twentieth and shift into twenty-first centuries (television, virtual systems, Web 2.0) primarily act upon mediatized spectators and change their perspectives in ways to match the technology's function. iDevices affect mediatized spectators similarly, but these same spectators also act upon the devices in ways that form a symbiotic relationship where no discernible dividing line between body and technology exists. Through these relationships, people are habituated to co-inhabit the spaces of the virtual and the actual simultaneously. Because of the technology, the iGen has entered a paradigm of sociality where actuality and virtuality can no longer be signified as separate or divided. Likewise, these mediatized spectators' perceptual apparatus and their digital environments become two and one. This double union creates a paradigm where perception becomes altered fundamentally, changing how they interact with the various modalities of performance and experience. The iGen constitutes its reality and likewise its embodied self based on a virtual/actual dialectic unfixed to any tangible physical location. The iGen is "the first generation to enter adolescence with smartphones already in their hands" (Twenge 2017, 5). We can even shift that idea of in their hands to really mean embedded within their embodied consciousness. It is a generation formed in the age of pervasive connectivity. Locations and spaces are never static in these mediatized spectators' perception of reality, hence disrupting an ability to detach from the ubiquitously mobile locations, let alone focus on one form of prescripted reality seen in static and unidirectional modes of storytelling.

In a nod to the participatory condition, Rosen (2010) refers to the iGen as a cohort of content creators through their constant process of uploads, posts, "likes," tweets, blogs, vlogs, etcetera (43). He argues this has led to a paradigm in which "They believe that they literally cannot perform only a single task at a time without being bored to death" (32). In his research, the average late-stage Millennial and/or early iGen student had between a 57 percent and 88 percent chance of multitasking with media during most activities, and a 73 percent likelihood of multitasking with some form of media during face-to-face conversations (82). Their total combined time multitasking with media was over 20 hours per day (29). According to a 2013 study backed by Facebook of nearly 8,000 18–44-year-olds, 84 percent of the time used with their smartphones was for non-voice applications such as text, web surfing, or social media (IDC Custom Solutions 2013). They have their phone on them for all but two hours per day. 49 percent of these respondents also stated that they were without their smartphones for less than 30 minutes per day, and 25 percent said they can't remember the last time they were without their phones. Of those in the iGen range, 74 percent stated the first thing they did upon waking up was reach for their phones. Today these numbers have only increased across the board, but the iGen were the earliest adopters. In one of the more recent studies, 48 percent of iGen members reported that they are online almost constantly (Perrin and Atske 2021). The numbers show that for the iGen, mobile devices have become their preferred way of communicating and connecting with the rest of the world (Poláková and Klímová 2019). With the influence of these devices being constantly present and connected, they live in a deeply mediatized world that is always immediately available and accessible via direct digital interaction brought about by pervasive connectivity. Today, the youngest iGen students have just started high school. They have known nothing but a world saturated with pervasive connectivity, one where their mobile devices have become part of their understanding of individual selfhood. They are forever intertwined with(in) the multiplicity of domains that make up the digital social sphere. They are literally walking, talking, sending, receiving, responding, and relating nodes within the vast digital network that is today's reality. Their entire perception of reality is based in a constantly on and connected model founded on digitally augmented relationality making them mediatized spectators.

Mediatized spectators' perception of the dual Reals in mobile augmented domains is now at the mercy of reflexive experiential interfacing, where the world, as perceived, is no longer a binary of static/dynamic inputs, but instead one made through its own perceiving inside of a perpetual feedback loop of interactions. Spectators raised in this paradigm simultaneously construct their reality and experience this reality through a reflexive referencing and reformation of the self, bracketing the networked world in which they exist. They can often feel a loss when the connection to the rest of the world is cut off. When encultured in a system of pervasive connectivity, that connection becomes part of one's personhood, and removing it has dire consequences. When an iDevice allows a constant and easily accessible connection to everything all at once, losing that connection (even momentarily) can create a troublesome rupture in one's sense of being. It is like experiencing a phenomenological form of digital phantom limb syndrome. As the first cohort born and raised in social worlds whose daily way of life is pervasively and inextricably connected through digitality, the iGen may be the first generation to live in an age where engagement such as what mobile technology allows is necessary. Multiplicity, fluidity, and flow, aided by digitality, guide their perception of the Real and shape their understanding of the social world. This generation operates in a perpetual liminal space between the virtual and the actual more than any before it. The touchscreen of the mobile device creates a proprioceptive link to the virtual, and through this link, it becomes part of the user's actual sense of being and selfhood. iDevices mediate and realize (perform) both the spatial configuration of place as well as the experience of space, time, and action. In essence, the screen spectators see the world through is augmented literally and figuratively by an adoption and addiction to the virtual gained through the use of iDevices. In 1996, Jean Baudrillard claimed that "we are threatened on all sides by interactivity" figuratively screened out by "video, interactive screens, multimedia, the Internet, and virtual reality" (2014 [1996], 192). Nearly thirty years later that outing has become a social norm enforcing cultural embodiment through the socially constructive influence of iDevices.

Game Play: The Gamification of Everyday Life

To best understand how the *Architecture of Game Play* operates as a system of/for experiential spectatorship linked to *pervasive connectivity*, it might be best to start with an understanding of games and gamification. Before we get to games, let's take a quick primer on play. Play is considered a form of activity separated from the mundane everyday life of human beings though it is uniquely a foundational part of our social life. It operates as a form of imaginary escape into a fictive reality where the rules of social existence can be rewritten and explored in ways that expand the possible. While around since the beginnings of human social formations, the conceptual theorization of play was first thoroughly explored by Johan Huizinga (1955) in *Homo Ludens*. For Huizinga, play was a form of escape from the socially dictated boundaries of real-life. Those boundaries were then replaced by a new symbolic boundary described as the magic circle. This new imaginary frame works to create "temporary worlds within the ordinary world, dedicated to the performance of an act apart" (10). These new worlds are then structured by their own set of formal rules, establishing the ludic space/time formulation. Inherent in the real world/ludic world dynamic is a sense of hybridity, as both exist in parallel and have the capacity to bleed into the other, impacting their fundamental construction. For the player in the ludic space, they are not always aware of the two overlapping realities, however. Gordon Calleja (2011) explains how these two realities are forever conjoined:

> Reality cannot be bracketed by closed or open circles, even if we could argue that such bracketing is logically possible. Reality does not *contain* play: like any other sociocultural construction, play is an intractable manifestation of reality. A consideration of games—whether from a perspective of the game as an object or as an activity, or the game's role in the wider community—*is* a consideration of reality. (48)

As such, play is an alternate reality that cannot ever be fully separated from the Real, in an equivalent manner that the virtual and the actual have been linked through the technic of pervasive connectivity. While once considered a separate activity, play has become increasingly embedded in all aspects of mediatized daily life.

Play theorist Roger Caillois (1963) pushed back against some of Huizinga's ideas related to rule-bound play based on structure to consider aspects of free play or *paidia*. Examples include the way a child might play with a toy simply for play's sake. The toy can be anything that fits the imagination of the player. Another way to think of this is to consider the way one might play an open-world video game today, like *Red Dead Redemption II* or *The Legend of Zelda: Breath of the Wild*. As a player within these digitally rendered alternate realities, one can simply wander about, explore, experience nature, or even start non-consequential fights, and collect plants and pelts simply for fun. This is free play, or play that is unstructured and non-goal oriented. This is true because these games allow near unrestricted exploration due to their open sandbox game design. Free play is the ultimate form of liminal activity, as it does not require advancement toward a goal per se, simply an experience of stasis in the playing space, in a manner like immersion. However, if the player wanted to instead win one of these games, they would have to engage in a variety of "missions" that help propel the narrative along, which then leads to an eventual outcome. Ludic play is therefore a form of purposeful participation where the player seeks to achieve some form of reward by engaging with the structure in a way that conforms to the system's rules, allowing forward progress.

In *Manifesto for a Ludic Century* (2015), games designer Eric Zimmerman argues that today there exists a paradigm of playfulness born out of digitalization consisting of "networks [that] are flexible and organic" (19). These networks are everywhere, and the iDevice operates as one of the primary mediators. The device links the network(s) to a person's perceptual apparatus as part of those networks and has the capacity to invoke a playful and gamified nature in the person's perspective of the social worlds they interact with(in). This playful condition encourages a perspective constantly in flux and moving toward new moments or possibilities, like the system of rewards inherent in gamification. Through the gamification of social life, mediatized spectators increase their capacity to engage in meaningful agency and choice making toward their respective goals, while following the rules of game play. The mediatized spectator's daily experience of life becomes one of a player who navigates the ludic spaces between virtual and actual Reals simultaneously. The way they interact with their mobile devices replicates the logics of game play, where navigating the social world requires a form of gamification.

At this point, I have mentioned gamification multiple times without specifically defining it. The term refers to the application of game mechanics and principles across various social and cultural domains. The basis of gamification is to use game design fundamentals in non-game related activities

(Deterding 2012). It is used as a way of motivating people toward specific outcomes and desired behaviors utilizing systems of challenges and rewards. It is also a process that purposefully modifies the experience of interacting with the mundane. Gamification within the digital social spheres we see today largely began as a way of social engineering consumers toward specific product use and purchase. Many see this as simply a form of digital manipulation. Because of the way it is linked to social media it has also been criticized as a tool addicting users to media platforms. Others see it as a way of making the world a better place (McGonigal 2011). The earliest notions of gamification within digital domains began in the early 2000s and were directly attributed to the mass deployment of digital games and digital systems, though the philosophy behind it existed for quite some time. Mobile platforms just became the ideal tool to harness and fine-tune its use. We see this occurring throughout the early 2000s as Web 2.0 systems began to focus increasingly on user engagement. The introduction of iDevices served as the ideal way to maintain near constant engagement through gamification, and many design thinking protocols for the apps run on these devices utilize game techniques. With the increase in iDevices deployed in society the rise in gamification began (Figure 4.1).

While gamification has become pervasive, it is a process that goes largely unnoticed due to the way it seamlessly layers onto nearly any interaction. For instance, in education, we have been using gamification for quite some time. Think back to your primary school days, where you might have been rewarded with a gold star for a particular activity, such as good

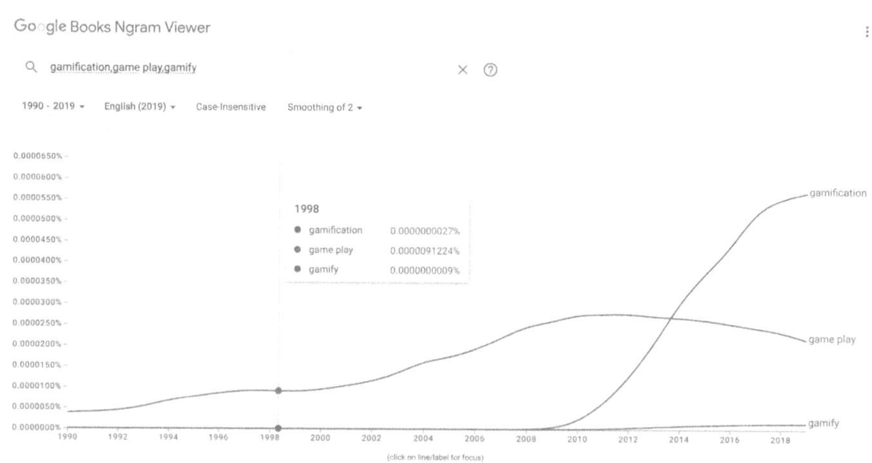

Figure 4.1 Rise in terminology related to games 1990–2019.

behavior during recess. Coming from a background where going out for dinner was rare, I remember being especially motivated by the Book It initiative in the 1980s, where we were rewarded with an individual Pizza Hut pan pizza for completing book reports from extra reading assignments.[7] This works due to the way games capitalize on human psychology that is geared toward fulfilling intrinsic and extrinsic rewards to maintain happiness (McGonigal 2011, 45). For me, the pizza was an extrinsic reward, a) because it represented a special event and b) because I liked pizza. For students today, instant gratification comes through the use of digital rewards in educational game platforms like Kahoot and Gimkit. Games push us toward fulfilling these reward systems as they strongly impact our neurological systems when we play them, whether we win or lose. Like any drug or other delivery system for neurochemical stimulation, games engage our release of cortical hormones such as dopamine that regulate our drives and desires based on emotional engagement and fulfillment. The introduction of gamification into daily digital interactions is simply a way of controlling our systems of engagement by subconsciously "gaming" motivation. This may sound nefarious, and can be, but gamification can actually have many positive impacts when harnessed toward meaningful experiences.

Gamification can motivate mediatized spectators through the process of meaning-making and making meaningful experiences out of playful interactions. Zimmerman (2015) states: "When information is put at play, game-like experiences replace linear media. Media and culture in the Ludic Century is increasingly systemic, modular, customizable, and participatory" (20). Zimmerman's paradigm is one where our technological interfaces shape culture into game-like systems, and gamification becomes the backbone of all interactivity. Culture predicated on the pervasive use of iDevices and the gamified operations these devices bring has the capacity to change the way one personally interacts with the world on a daily basis. When a player learns to understand—either consciously or subconsciously—how their actions impact the structure of a game-world and, likewise, their everyday experiences, they enact forms of meaningful experience. I've argued that participation and immersion impact the expectations of the spectator, game play ramps up those expectations by actively rewarding the player for their interaction in meaningful ways.

Taking a sociological and performance studies perspective on the relationship of play and games to society, McGonigal (2011) has argued that all reality has been broken in the twenty-first century and that the gamification of that reality is one way to fix it. She points to a prevailing condition where gamification is becoming an underlying framework for dealing with everyday life in the early twenty-first century. She argues that through games and game structures society has the capacity to address large-scale problems.

Working through various play theorists (Abt 1970; Caillois 1963; Costikyan 2002; Huizinga 1955; Suits 1990) McGonigal argues that game play is differentiated from other forms of play based on four basic traits. It must contain: (1) a goal; (2) rules; (3) a feedback system; and (4) voluntary participation (21). Sets of rules are what make game play different from the free play described above. Katie Salen and Zimmerman's (2004) contribution to the discussion on game design adds two additional traits: (1) conflict and (2) a player (80). Each of these requirements is clearly embedded in nearly all interactions one has with their iDevices, save one, volunteer participation. I propose that the constant connection to iDevices complicates the idea of voluntary participation argued for by McGonigal. Of the play theorists tracked by both McGonigal, Salen, and Zimmerman, only Caillois (1963), Suits (1990), and Sutton-Smith (1971) agree that voluntary participation is necessary. Considering how all life becomes gamified via connections to iDevices and playable media, it is hard to argue that voluntary participation is part of the equation. In today's technologically connected ecosystem, smartphones and mobile media become part of a human being's subjective system of perception and social world construction and, therefore, are no longer part of a voluntary condition. This helps explain why disconnecting is so difficult.

When considering the ways that iDevices utilize and promote game play, one must simply look at the interactions that they are built around. Mediatized spectators have become conditioned via iDevice technology to receive rewards in the form of notifications, emails, tweets, retweets, pokes, likes, virtual rewards, and such. These interactions are more than *paidic*; they are ludic in the way that they structure one toward specific micro achievements. As Calleja (2011) explains, these ludic structures "appeal to our tendency to organize the complexity of our lives into hierarchies of goals by creating systems in which these goals are explicitly stated and specifically designed to be attainable" (150). A mediatized spectator performing as a player follows these specific sets of guidelines to achieve desired goals with(in) the performance mirroring the playful formulation of life under pervasive connectivity and gamification. Within the frame of daily life, the connection to an iDevice increases one's desire to achieve definable goals. For example, the response enacted when constantly checking our smartphones for incoming information acts as a form of instant gratification that we are ingrained to constantly pursue. That constant seeking of digital gratification begins to impact the intrinsic wants and needs of the user by requiring more possibilities for rewards deliverable both by the device and beyond the device. Because of their portability and connectivity, these devices implement game logics with every use and disrupt previous social norms and expectations toward following the operations and aesthetics of digital games. iDevices not only act as a technological augmentation between place, time, and space, they also operate as ever-present connective systems of objectives and rewards in a game-like manner. This system then maps onto a mediatized spectator's entire understanding of reality, or as mobile media researchers Hjorth and Richardson (2014) explain, "ludic experience leaks into the spatial, temporal, social, and corporeal affordances of everyday life" (25). Via the term nomophobia—where a user of virtual technologies becomes figuratively and literally addicted to the immediate access and response available via the technological interface—gamification capitalizes on the neurological subjectivizing capabilities of iDevice technology. This ever-present connection augments a person's sense of agency, overlapping the affects and effects found in the previous architectures and technologies. The relational connections to pervasive and mobile smart devices augment human interaction with(in) the world with game-like qualities and subsequently allow spectatorship to evolve into interactive modes of playing based on the structures of game play.

Game-play-modeled performance relies on systems thinking as a model for structuring experience. Systems thinking refers to a complex mode of conceptualization and analysis where one expands their viewpoint to focus on entire systems of interactions and interactors, even when individual aspects of the system are the object of analysis. It is defined as "a set of synergistic analytic skills used to improve the capability of identifying and understanding systems, predicting their behaviors, and devising modifications to them in order to produce desired effects" (Arnold and Wade 2015, 675). To think systematically means adjusting one's focus "away from causal paradigms and towards a perspective or approach that considers the interrelatedness of things" (Díaz 2019, 241). For example, consider looking at **both** the forest **and** the trees, instead of one preceding the other, as a useful way of capturing this way of thinking. In a system, there are multiple integrated pieces that interact in ways designed for optimal success. A game system, for example, is one encompassed by a specific environment that allows dictated forms of interactions. The game environment contains objects that may include human and non-human actors. These objects all perform as players in a relational process in which each has characteristics or attributes dictated by the rules of the game and the environmental constraints of the play taking place with(in) the system. The rules put in place determine how the players may play the game while also defining the basic relationships among all actors present. Game structures are more defined than other forms of play, but because of this, they also have the increased potential for meaningful interactions. When interactions are designed with meaningful experience in mind, it allows for the "perception of great 'game play'–a sweet feeling of being part of something large that is moving forward (the game's narrative) while you still get to steer (the game's play)" (Kelly 2016, 230, quotations in original). The player has the agency to impact their overall experience, but is given enough guidance, via rules, to not overtake the system's design, which allows the proper amount of tension between both the player and the structure.

Farman (2021) cites 2001 and 2002 as the years in which video games first outsold movie box-offices, first in the US and then globally. This shift marks a moment where interactive experience began to supplant so call passive reception as a mass market form of media entertainment. In 2023, the global video gaming industry had revenues of 180 billion dollars with an estimated growth to 205 billion in the next two years (Wijman 2024). The operative function of video games and their technogenetic influence ported to iDevices in the later part of the 2000s. Since the screens of most of these devices also require touch-based directions, users physically interact with the narratives and performances they consume in ways not seen before. These devices become figurative game controllers for the construction of reality. Every physical move of their fingers and hands brings them into direct contact with media. This connection is more "real" because there is a higher level of contact and interaction with one's perceptual apparatus. The haptic connection brings the user closer to the virtual event on the device. This proximity causes the actual and virtual to have moments of overlap. In an earlier article detailing the problematic nature of discussing violent video games as a precursor to real world violence, Farman (2010) explains that the shift in perspectives between the two Reals—virtual and actual or digital and analog—is at the heart of how gamification works in mediatized paradigms of digitalization. He starts by questioning Huizinga's dictum that play must operate in a sacred space outside of the Real. Farman argues that this is problematic because the sacred is "becoming less and less a part of game play, as seen in many digital games that take place within real-world space and intersect with real-world events" (101). In the liminal spaces established in a gamification of social worlds, alternate realities become present and accessible. In this in-between space, the possibility for critical discourse becomes available, but only when the architecture allows rhetorical strategies for questioning the position of the player as a flexible observer and participant with(in) the multiple Reals.

Performances that capitalize on the multiplicity of Reals often take the form of Alternative Reality Games or ARGs. An ARG—sometimes referred to as a mixed reality or augmented reality game—is a framework for performance that blends fictitious narratives with real-world spaces to allow the player the ability to navigate the in-between in a manner that gives some semblance of immersion while highlighting a sense of Brechtian alienation. An early whitepaper developed in conjunction with the International Game Developers Association defines an ARG in this manner:

Alternate Reality Games take the substance of everyday life and weave it into narratives that layer additional meaning, depth, and interaction upon the real world. The contents of these narratives constantly intersect with actuality, but play fast and loose with fact, sometimes departing entirely from the actual or grossly warping it—yet remain inescapably interwoven. Twenty-four hours a day, seven days a week, everyone in the country can access these narratives through every available medium—at home, in the office, on the phones; in words, in images, in sound. Modern society contains many managed narratives relating to everything from celebrity marriages to brands to political parties, which are constantly disseminated through all media for our perusal, but ARGs turn these into interactive games. Generally, the enabling condition to is technology, with the internet and modern cheap communication making such interactivity affordable for the game developers. It's the kind of thing that societies have been doing for thousands of years, but more so. Much more so.

(Martin, Thompson and Chatfield 2006).

The prevailing aspect of ARGs, whether they include various forms of technology or not, is the way they blend the ludic realm with the real world in a manner that blurs the distinction between the two, creating a hybrid reality. Lindsay Brandon Hunter (2021) describes ARGs as forms of pervasive games "that do not respect the magic circle" as "they lack a clear, stable boundary marking the boundaries of the place where (and sometimes the time when) they are played" (86). These are ideal hybrid spaces. Hunter goes on to explain how the construction of ARGs often focuses on making visible the boundary as a hallmark that creates a meaningful effect for the spectator:

> The pleasurable creepiness of ARGs is anchored in their refusal to acknowledge their own fictionality, but it would be impossible—and probably less fun—for the assets and modalities (websites, emails, and text messages) that comprise the fictional worlds of ARGs to be actually indistinguishable from serious ones. ARGs are best characterized not as wholly successful simulations, but as extravagant masquerades that never entirely succeed at effacing their own construction, marked more by a playful desire to disguise their strings and seams than by absolute success in effacing them. (96)

In the most successful ARGs, a player becomes part of both the narrative (virtual) and landscape (actual), while riding a liminal fence between the two through the mechanics of gamification. This "becoming part" gives the player a double presence, that is both inside and outside, and allows for heightened levels of critical awareness. When spectators gain understanding of how their actions integrate into the game's structure and how that integratedness is also applicable to the world beyond the game, they develop a critical awareness of both Reals. Meaning made with(in) the structure is also applicable outside the structure.

ARGs have been deployed in various formats divorced from mobile technology over the past few decades, but the most famous utilize the game structures of digital tech. Even so, Hunter explains that their ability to create a sense of the real in gameplay is partially based on their ability to capitalize on "digital networkedness, effectively rooting the games in the actuality of ubiquitous computing even while sketching the fantasy of an alternate reality" (95). The most famous examples began to arise at the turn of the century as digitalization first began to embed itself in society. Some examples include *The Beast* connected to the 2001 Stephen Spielberg movie *A.I: Artificial Intelligence* and *I Love Bees* in 2004 connected to the release of the game *Halo 2* for Microsoft's Xbox console system. Both Hunter and McGonigal cover these examples in more detail. For those looking for a more analog example of an ARG that has made it into the mainstream, *The Jejune Institute* played across various locations in San Francisco and Oakland from 2008 to 2011 and was adapted into the AMC television series *Dispatches from Elsewhere*.[8] The four-part durational narrative of the game pitted local players against the fictitious Jejune Institute, a global pseudo-science foundation led by the nefarious Octavio Coleman Esquire, formed with the goal of developing technologies that expand our awareness of the world around us. In each phase of the game, the players explored the city as a real-world treasure hunt with a variety of performance installations and encounters to find hidden clues hinting toward a magical other world named Elsewhere; a powerful place hidden just beyond our conscious perception of reality. Like many serious games that hope to lead players toward meaningful interactions, the game maker's final objective was to open the players' eyes to the beauty of the world around them. Jeff Hull and Nonchalance, the company behind the ARG, have continued to develop extensions of the project, each seemingly attempting to develop community in a world that has become detached due to the pervasive aspects of their mobile tech. All of these examples require extensive time to grasp the scope of an ARG. For those who want to understand one in action in a single afternoon you can watch David Fincher's 1997 film *The Game*. The lead character, played by Michael Douglas, is led through a series of near fatal encounters that, unbeknownst to him, have been staged by a team of professional "experience makers" as a way of raising his own awareness of his positionality as a member of an elite class. It is the life and death consequences of the game that become so meaningful. Like the lead character of the film, the line between fact and fiction becomes blurred for the player and invokes Richard Schechner's (2017) theorization of "playing in the dark," a form of serious play where the player is not aware they are playing a game. In the contemporary era of pervasive connectivity, one might say we are all playing in the dark, as our iDevices are constantly driving us toward unexpected goals and outcomes. Going even further than not simply knowing you are playing is the concept of "ambient play" (Hjorth and Richardson 2014; 2020) brought about by mobile devices, where the entire background of life becomes a game due to the pervasive connection to locative media.

The above examples are cited to help frame how ARGs capitalize on hybrid realities and show that they are not fully reliant on any specific form of tech. However, in the era of pervasive connectivity, the iDevice has helped to bolster the feeling of realness for the player. The sense of realness is often most felt when the purpose of the game is to address some form of issue to solve in the player's daily life or societal frame. When performance makers build game structures around social or political issues from outside the game-world, players gain an ability to critically deconstruct the affect and effect of the narrative on the Real (actual) world by slipping into the Real (virtual) of the game-space. McGonigal (2011) argues that the best ARGs often have audacious goals that implicate the player in narratives involving "entire communities or society at large" (125) and are designed to help make our social condition better and part of a bigger picture. It is precisely the act of play that allows them to gain a critical sense of how to game the systems of reality in order to solve the problems at hand. The critical nature of the player's interactions affects their understanding of their agency in the world beyond the game.

4.1   The Playing Spectator: iPerformance and Ludic Criticality

Through the Instances in this hinge section, I discuss game play experiences that utilize iDevice technology in various ways. Two are constructed as ARGs, which allow the player agency to engage in ludic critical exchange between two possible Reals. The focus is on the experience of the player who engages with ludic criticality through the various ways they play within the game structures. For the spectator as player, the relationship between the gamified experience and the worlds they play with(in) expands to create new perceptions of time, place, and space. This experience offers them a critical and ethical positioning while also allowing them to question their very own acts of play within the experiential system of exchange. In the last Instance, I discuss my own experience of a playthrough of a game that brings to the forefront the critical perspectives of the use of the iDevice itself. This mode of playing multiplies the affect of the hybrid positionality as it causes me to not only question the reality of the game, but also my own implicitness in the ways gamification impacts those in spaces outside my immediate domain. Through each, the player's experience is framed by the impacts of pervasive connectivity.

Instance: (Re)Contextualizing Time, Space, and History with
Pocket Monsters

I am walking around the historic district of downtown Santa Fe, New Mexico, on a brisk January afternoon. The sun is high in the sky and is causing reflections on my iPhone which I'm using to navigate the square around the

Santa Fe Cathedral. The square and the cathedral have multiplied and taken on manifold identities this fine day. They exist as cultural landmarks that I can explore physically, giving me the historic background concerning the formation of the southwest city, but they are also digital landmarks, marking a stopping place for engagement with virtuality. The Santa Fe Cathedral Monument stands about ten feet in front of me. It consists of an intricate menagerie of domesticated animals and humans frozen in bronze (Figure 4.2). At the base are a donkey, a sheep, a goat, a pig, a rooster, and an ox. Atop are a Spanish soldier, a monk, and a peasant family consisting of a woman, man, and two small children, one holding a doll. Hanging in the liminal space between human and animal is a cornucopia of cultural riches that connect these creatures to the history of the town: Fruits and vegetables, musical instruments, a Bible, an axe, and various weapons. Above these figures is a miniature figure of Mary La Conquistadora (Our Lady of the Conquest), a statue representing the Virgin Mary who guides and anchors these figures as part of a unique 400-year history of the Santa Fe Cathedral and the subsequent city that formed around it. While reading the plaque (Figure 4.3) embedded in the monument's base, I receive a notification from the iDevice in my right hand. I look at it and see that there is another figure perched nearby the monument. It is a figure that for all senses and purposes is invisible in the actual world where the monument sits, but with the simple tap of my left finger, the creature now shares this space with me and the sculptured figures. I have found a Houndour (Figures 4.4 and 4.5) sitting at the base directly across from the ox. As I move my phone around in my hand, the virtual creature moves with me, adjusting its position to sit now on top of the statue as a digital overlay. Before I can catch the creature with one of my PokéBalls or get a good image of it "in the wild," it runs away, and I am left wondering what other creatures might lurk around the next corner. I look down at my virtual map and see a Hoothoot at the central square (Figure 4.6) just waiting to be caught; after all, the goal is to "catch them all." To do so, I must traverse the actual space of the historic square while also crossing over into the virtual space of the app-based game titled *Pokémon GO* (2016).

Launched in 2016, *Pokémon GO* is an AR game that plays on nostalgia of the 1990s and early 2000s when the first Pokémon (Pocket Monster) craze took shape through its introduction as a Game Boy hand-held video game. Using AR technology that creates a digital overlay on the image processed by your smartphone's camera, mystical creatures appear in front of your eyes as if they are actually in the space but only made visible by using special glasses. The AR and the GPS (Global Positioning System) software on my device operate in conjunction with the game's architecture to populate these creatures and to move them through actual space with me in real time as I travel around the world. My purpose in the game is to capture and train a collection of these virtual creatures as they lurk digitally beneath the quotidian features of our actual world. By loading the game onto my smartphone, I have created a link between the virtual game-world and the actual physical world

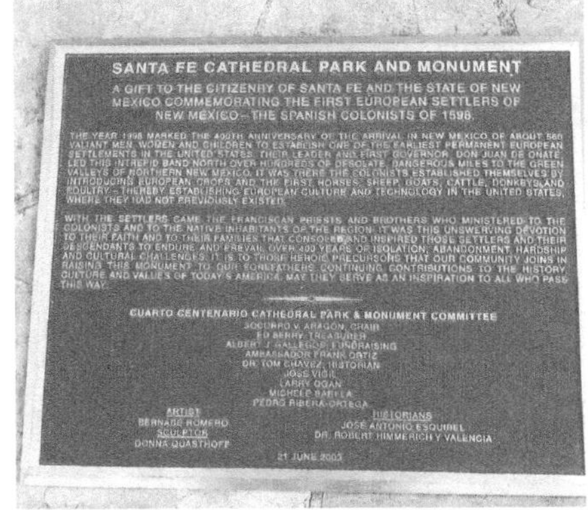

Figure 4.2 The Cathedral Park monument, Sante Fe, New Mexico.

Figure 4.3 The Cathedral Park monument plaque.

based on landmarks, human interactions, and geographic realities. These landmarks are geotagged by previous users of an earlier Niantic game and act as PokéStops within the virtual realm. The stops serve two purposes: As virtual guide-points attached to my digital map where my avatar (Professor-Cobrah) gains items needed to catch creatures, and as information collection points about the specific landmarks as they exist in the actual world. Like the digital creatures interacted with in AR mode, these PokéStops operate as limens between two Reals: The virtual and the actual. My job as a Pokémon hunter is to navigate these two spaces in a ludic journey, to find, catch, train, learn, and explore.

I take a few minutes to walk around the statue, read the inscription, and learn about the founding of the city and its 400-year history while looking at the figures on the statue. These artistic renderings transport me to a different time and place, one in which the founders of this physical location had to endure a multitude of difficulties to establish their way of life. The statue exists as a monument to those endeavors. As a piece of art, it urges me to travel through multiple time frames in the virtual space of my mind, imagining what it must have been like to live in this arid and mountainous climate as a settler of a harsh new land with unlimited possibilities. In a similar manner, the device in my hand allows me to travel into another virtual space intricately connected to this historic place. By attempting to catch a Houndour, Pidgey, Hoothoot, or

Figure 4.4 Screenshot of Houndour on digital background *in Pokémon GO.*

Figure 4.5 Screenshot of Houndour seen through an example AR overlay inside *Pokémon GO.*

Snorlax, the game encourages me to interact with the physical space in ways I would have ignored had it not been for the overlapping realities.

I walk over about one hundred yards to where I saw the Hoothoot on my map and encounter another physical monument to the city (Figure 4.7). This time it is a forty-foot-high obelisk memorializing the war heroes "who have fallen in various battles with Indians in the Territory of New Mexico." I find it apropos that this monument also serves as a Gym inside the virtual space of the game. This PokéStop operates as both an actual and virtual marker for space where historic and game-based land wars occur(ed). Gyms are unique PokéStops that serve as collection and drop points for trained Pokémon. In these virtual arenas, other Pokémon from my team[9] can battle creatures from other teams as a way of marking the ownership of virtual territory (Figure 4.8). By holding the gym through battles, my creatures can level-up, and I gain PokéCoins, the virtual currency used in the game. When entering the Gym, I see other creatures ready to hold space, but I can also tap on an icon that gives me information about the physical landmark upon which this virtual battle arena sits. Before battling, I take the time to read about the monument. I learn about the defeat of the Federal Army at the

Battle of Valverde, fought in 1862 (Figure 4.9). This information is a digital version of the inscription on the actual monument towering above me. I move back a screen and am again with my team ready to join the fight, hoping that I won't fall like those that came over a hundred years before me. In the virtual realm, I interact with the actual space in a form of ludic action where I symbolically re-enact the historic accounts of real-life warriors and victims. This act of game play makes me more critically aware of the history of this physical space by allowing me to virtually embody the actions that took place in another temporality. While not every example of a *Pokémon GO* Gym carries with it this specific historical symbolism, the fact that it exits here on this physical landmark allows me to engage in a mode of ludic criticality, where I consider the relationship between multiple spaces experienced through my game-based actions. The use of AR on my iDevice creates linkages between the digital, the corporeal, the real, the fictive, the historic, and the now. The device augments my sense of place and space while also altering how I perceive personal selfhood in relation to temporality and historicity. Here, in historic Santa Fe Plaza, *Pokémon GO* allows me to traverse these multiple realms in a ludic adventure through the multiple times and the multiple places that connect the virtual and the actual.

Figure 4.6 Hoothoot at SE corner of Santa Fe Plaza Square inside *Pokémon GO.*

Figure 4.7 PokéStop/Gym – *To the Heroes* Obelisk in virtual space in *Pokémon GO.*

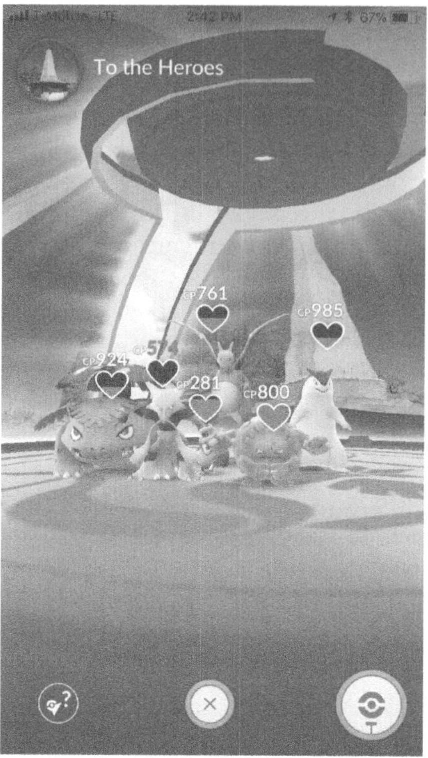

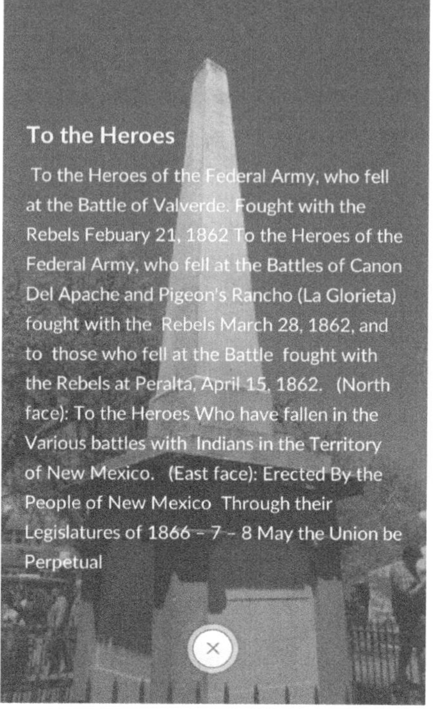

Figure 4.8 Team Mystic ready for battle in *Pokémon GO.*

Figure 4.9 Digital readout of monument inscription in *Pokémon GO.*

Instance: Adventure One – Traversing Urban Liminality

In 2015, the London-based performance collective Coney first produced an ARG-like performance titled *Adventure One.*[10] The locative-media based game/performance used digital augmentation and spectator interaction to draw players into semi-personal narratives that require them to employ *ludic critical agency.* This form of agency develops out of the interactions between the affective and tangible forms of agency discussed in the previous two chapters. Interaction between these multiple agencies allows the player increased ethical introspection and reflexivity. Ludic critical agency blurs the line between the actual and the virtual and introduces an ethico-political tint to the performance experience via the ludic frame. In *Adventure One,* the player accesses and employs this mode of agency by using an iDevice interface, which enables new possibilities and understandings of proximity between places (local/symbolic) and spaces (actual/virtual). In *Adventure One,* a spectator takes on the role of player, adopting both the aesthetic condition of immersion and the political capacity of participation.

Coney integrates media artifacts and logics into many of their performances. Their work is often covertly political while maintaining a level of intimate playfulness. Often, a Coney performance starts months before the physical ticketed event. The collective often initiates gameplay for the spectator via email or text message as a figurative rabbit hole to drop down into and explore, much like a twenty-first-century Alice. The beginnings of the group are a playful mystery, reportedly having been initiated by the enigmatic and elusive Rabbit; a figure that pops in and out of the group's narratives and playmaking. Per Tassos Stevens, the collective's primary playmaker, Rabbit institutes the "Loveliness Principle" at the heart of all the group's performance work. Coney describes loveliness as a participatory performance that "play[s] with ideas that resonate in the world around us: From the everyday to the extraordinary. Our work is inspired by the belief that the world can be a magical place where ordinary people can do extraordinary things" (Coney n.d.). This mode of play often incorporates mediatization of social and performance narratives. When interviewed by Josephine Machon (2013) about the use of media, Stevens responds, "for us, the important thing about it is that it comes from a principle of gluing together a multi-layered audience experience," (199) and continues by stating that their goal is "making stories happen in a sense of a journey. I like the idea that we're making a world, or a lens for the real world" (201). Critics often lump Coney's work into the growing canon of immersive theatre companies; however, performance works such as *Adventure One* show how playing-theatre based in structures of gameplay is a more apt term.

Coney designed *Adventure One*'s primary trajectory around a narrative concerning industrial espionage and algorithmic technologies that have the capacity to shake the very foundations of world financial markets. As a player in the game, the spectator is initiated early on as a potential spy and tasked with following a set of clues placed in real-world settings that lead them to confront an ethical dilemma concerning the proper use of certain technologies. The performance's settings are undisclosed to the player until the day before the date of the ticketed event. The only advance information given is that the event will take place somewhere in London's financial district. Once inside the physical playing space, the spectator is led from a telephone booth to a convenience store, to a cathedral, to a business plaza, and eventually to a local pub. Each stop on the journey, and the spaces between each stop, become crucial physical landmarks where the player learns more about the inner workings of world finance and this world-shaping industry's connection to historical landmarks in the City of London.

One crucial way *Adventure One* differs from other experiential performances is the use of its real-world settings. As Adam Alston explains

(2016), "the very notion of staging reality in immersive theatre tends, more often than not, to be avoided by immersive theatre makers who strive to achieve ever-more total closure of a fictive cosmos" (61). *Adventure One* expands the fictive cosmos beyond immersion by intermingling real-world locations with virtual and actual spaces, and virtual and actual storylines. The night before the event, the player receives a digital map via email to load onto their iDevice and a set of audio tracks to listen to that correlate with physical locations. The iDevice helps to further develop the sense of gamification in the player by becoming a crucial tool to access and assess the performance network which is made through the link to multiple realities.

The spectator also connects to a digital proxy/avatar the moment they sign up for a ticket by receiving an email that instructs them to await a call. On the other end is a recorded operator named Josh who asks if the player is willing to take responsibility for their actions in the upcoming performance. Answering no terminates the interaction. Per Stevens, answering yes leads to another series of questions gauging the player's political and ethical expectations and understanding of the global financial system. These questions implicate the player as an integral part of the upcoming narrative based on the agency to take part and the willingness to communicate with(in) the structure. The iDevice also works as a conduit to connect the player in the weeks leading up to the event and eventually leads the player to a previously unknown location where it helps guide the player through a cat-and-mouse game with live individuals; individuals who are never explicitly explained as actors or just regular people until the final moments of the performance. Stevens explained to me that the smartphone also becomes a "prop that enables them [the player] to blend in during their covert mission." The experience capitalizes on the player's use of iDevice to help track the player's movement through physical space while correlating that placement to the locations on the virtual map. The first connection to Josh as a narrative proxy on the device creates a relational aspect between the player and the game weeks before the event, allowing for a deeper sense of agency and involvement in the full shape of the event. The proxy continues to operate via an algorithmic program embedded in various locative media channels that the player accesses along the journey through the recordings, voiced by either a new character Fiona, or the original contact Josh. As there is no internal character dialogue other than guided audio tracks, the digital proxy works to engage the spectator as a player in a game that only progresses through the connection between the real-world space, digital information, and the player's interactions between the two. The player connects actual and virtual world information through the digital proxy, using various texts, phone calls, and emails sent and received via smartphone or other iDevice.

To create interactivity with the player, Josh and Fiona connect deliver text keywords connected to the downloaded audio playback. These technological co-players create connective tissue between the player and the event's narrative unfolding. The interactions allow a deeper form of agency and involvement in the event based on the player's direct input, creating a feedback loop of meaning-making via technologically augmented embodied space. As a locative game, the mobile device allows the physical spaces to be "layered with other worlds, and the full sensory-inscribed experience of these spaces depends on successfully navigating the permeable delineation between them" (Farman 2021, 78). The program and the device augment not only the narrative and the environments where the narrative occurs, but also the player's subjective position with(in) the event because the plot trajectory requires navigating the space between both the physical and digital information accessed.

By facilitating the sending of text messages, answering of phone calls, reading of digital maps, and listening to recorded audio, the iDevice replicates not only the embedded nature of mediatized sociality but also acts as a user-initiated locative tool in the real-world game setting. As a version of locative media, the iDevice is used in the performance narrative to "generate new potentialities for facilitating the forms of social appropriation, citizenship and (experimental) sociability" (Wilken and Goggin 2015b, 5). The proprioceptive link to Josh and Fiona also allows the player to discover new information about the site-specific location of the event. This information adds to the critical awareness about the finance industry the game intends to question. Midway through the game, as the player is wandering among the mix of historic and contemporary structures housing the inner workings of global finance, Fiona instructs the player to:

> Look at the buildings around you. The architecture is trying to tell you that this is the heart of the city. But it's a façade, a collection of fronts and hidden levels … Who owns these buildings, do you reckon? What secrets lie behind their doors? What secret levels beneath your feet? The markets these days live in a secret location … The market is a computer server, a data centre, somewhere a mile from here in a top-secret location. And most of the traders are algorithms.
>
> (Coney, Will Drew and Tassos Stevens,
> *Adventure One*, 2015.)

The player's proximity to the physical sites calls attention to the political implications of playing the game. The mediated interactions with Josh and Fiona help develop a sense of relational proximity by allowing the player to connect with supposed real people, pushing the player on to success. The player's task is to test different levels and modes of agency through technologically guided interaction. For example, at one point Fiona asks the player (via audio track) whether they would like to embody the actions of the antagonist of the story, by re-enacting one of this character's daily activities: Lighting candles in a local chapel. This moment of reflexive agency allows the player to embody a foe, forcing the player to critically evaluate the ethics of doing this act in a real-world environment. The player must balance the liminal space between the fictive world created and the actual world framing the fiction.

The player correlates the locations interacted with(in) against the maps accessed via their iDevice. By listening to pre-recorded audio tracks and sending/receiving text messages with the digital proxy, the player fills in the gaps between the information present in the real-world setting and the game's construct. Digital assistance is necessary to advance through the narrative, with each step in the adventure unfolding as a live-action video game coaxed on by the virtual guide. When asked about the inclusion of the iDevice and its relation to everyday life, Steven's explains that it allows "given permission to temporarily suspend the rules of the game" at any time. Her goes on to argue this ability to pause the game and the fictional reality while still attending to the device allows the player heightened levels of agency. Because the player is in constant interaction with the iDevice, they replicate the daily grind of a typical twenty-first-century city dweller and can, therefore, maintain their covert status while taking a breather. The device becomes their cover.

One of the overarching goals of the experience is to highlight the player's location in the physical environment as a way of affecting their perception of the entire financial system. One might see the buildings and topography making up the London financial district as a quotidian landscape until virtual and corporeal augmentation interrupt the daily narrative, allowing the player to perceive the space in a critically affected manner. In conjunction with the device, the player reorients their perception of the location through a mode of interlinking performativity. The connection to virtual maps, information, and avatars allows a complex multiplicity to arise in the way the game unfolds and offers modified notions of proximity with(in) the performance framework. The way the player engages with the city is similar to Rowan Wilken's (2014) description of the epistemological relationship between locative media and narratives of proximity and space:

> It is a space in which it is possible for an urban dweller to take pleasure in being drawn out of oneself. To approach the city in this way is to understand that other meanings, practices, and perspectives on the city are possible and which can lead to opportunities for learning and new or different experiences. (178)

The conceit and concept of the production is to change the perspective of the charged space for the player. iDevice interaction allows the player to traverse the liminal spaces, engaging in a mode of ludic criticality. This type of exchange mirrors a contemporary paradigm where "the process of inhabiting multiple spaces simultaneously has moved into the sphere of the quotidian and often goes unnoticed" (Farman 2021, 87). The saturation of information and communication technologies in deeply mediatized societies allows a hybrid subjectivity, through which spectators habitually negotiate the in-between spaces created at the intersection of the virtual and the actual as a process of bonding. When this bonding occurs between the two, each space develops an inseparable connection to the other and allows bi-directional fluidity. Because of the pervasive influence of iDevice technologies that constantly travel with us, the virtual and the actual are inextricably bonded together, which makes the two indistinguishable from each other. The player can then critically evaluate the way they participate with(in) the location and narrative, which subsequently increases the impact of agency through the interactive exchange. Navigating meaning-making in real-world locative media games such as *Adventure One* replicates this bi-directional fluidity and harnesses its potential.

Following Coney's goal of making the world a lovelier place, the player is effected and affected in a manner that asks them to consider action against the proposed negative forces of capitalism that the financial district metaphorically represents. No longer is the playing space a common arrangement of buildings and commercial enterprises; it becomes a network of loaded signifiers pointing to systems of capital: Systems to actively question. The climax of the game is a theft, one relevant due to the recent near collapse of the global economy. The player, who has been led through various tasks to level up their ability for critical reflection, is asked to steal a briefcase from an unknown man in front of one of the buildings. This act happens in broad daylight, outside the confines of a contained scenography. Submitting this task could very well lead to others in the environment taking negative action against the player. These passive watchers have no clue that a "performance" is taking place: All they see is a potential felon stealing a briefcase from a businessman who may work in the building in front of them. A critical function in this moment's success is the realization that there is reflexive relationality between the real-world pedestrians and the fictive-world players. The request to become a thief reinforces the bonding and blurring of the virtual and actual, allowing the player to engage in ludic critical exchange. As exciting or appalling as this act might be for the player, the theft is not the most critical moment of the performance in terms of real-world impact. After the successful theft, multiple players assemble in a local pub to decide what they will do with the stolen briefcase. The briefcase contains a disk with an algorithm that can manipulate the stock market and allow the user to gain riches beyond their wildest dreams; however, this will also upend the social dynamics that the current financial system is built upon. Though framed inside the narrative, the question has symbolic weight in the actual world the players have now re-entered. This final "scene" straddles fiction and reality and acts as a debriefing session with a lasting moment of critical reflection. This debrief is crucial to the agency transmitted in and through the performance. Through the debrief, the players disconnect from the narrative, throwing off the ludic veil to critically engage with the material as non-players, activating continued political and social agency beyond the event.

The unique connection created by the tethering of device, player, and multiple interconnected spaces creates a form of exchange based in ludic criticality. The experience is meaningful because of this exchange. In an essay about a different performance by Coney, Gareth White (2016) explains that in the experience of being a player, one feels "a responsibility for solving the problem, evading its traps, presenting a better solution than the dilemma ostensibly allows" (22). This feeling of responsibility inside the narrative construct is one way in which the rules of game play mark the difference between a simple immersant, participant, or player. The player assessing their own responsibility allows for a dualistic viewpoint between the narrative event and the real-world constraints. The player must maintain a constant connection to a rational and conscious duality in the event as a critically affected spectator of the event. Through this dual consciousness, the player gains a heightened potential for instructive and social agency. This agency is highlighted when performance makers structure their games around political motivations the way *Adventure One* does.

Instance: Phone Story – Ludic Criticality and Ethics Under Pervasive Connectivity

I end this section by addressing a trait of gamified spectatorship dealing with ludic criticality, namely ethics. If part of a mediatized paradigm is to bring into relation the various technologies, actions, and agencies of the spectator, then it is necessary to not simply extol the virtues of the iDevice as part of a techno-utopian understanding of the world. As stated before, iDevices have become a part of a human being's sense of their self and their place in the world. Connections to the devices augment the way performance is perceived and how spectatorship can be used to reconfigure perceptions of location. This is readily experienced in *Adventure One*. The reconfiguration of perception is in constant negotiation when human beings enter the in-between spaces of play that iDevices initiate. In this liminal space, there is a unique capacity for performance and games to open up critical moments of reflection. Mary Flanagan (2013) characterizes critical play as "a careful examination of social, cultural, political, or even personal themes that function as alternates to popular play spaces" (6). Flanagan argues that games are themselves a social technology that, when used for critical purposes, can provide new lenses for looking at the work controlled by analytical and investigatory frameworks (6–7). Games are designed with a specific subjectivity in mind, and when they use criticality as both a goal and tool are designed to allow diverse perspectives toward developing multiple solutions to real-world problems (261). A ludic critical frame for approaching the world is one that engages in multiple questions and multiple perspectives to understand the various potentials toward multiple futures. In this way, ludic critical exchange is also a mode where a player can enter into the game performance to look anew at a subject such as location or even the very object that delivers the game they are playing as a form of ethical encounter and questioning. *Phone Story* is one such example.

Produced by the activist games company Molleindustria, *Phone Story* (2011)[11] is an app-based mobile phone game that uses ludic criticality to problematize the very platform and device one plays the game on. The gameplay and narrative ask its player to question the ethical underpinnings of the creation of the device itself during the game's play. The game's structure is based on four mini games that implicate the player in the unique global process of smartphone creation; from the extraction of its material parts to the labor practices that support its production, to the marketing and design strategy that make it an indispensable cultural product, and finally its impact on the world as a tool designed for planned obsolescence.

Screen 1:

Hello consumer. (Smiley Face) Thank you for joining us. Let me tell you the story of this phone while I provide you with quality entertainment.

Upon starting the game, I am greeted by a talking digital face, ostensibly the face of the very smartphone I am holding in my hand. This face reminds me that my deviceis more than a technological object; it has a life and agency much like myself. I am reminded that it is important to understand where it came from and why my use and connection to it forces a larger question about my place in the world and existence. The smartphone is asking me to enter the liminal game space, knowing that I will be looking from the outside while engulfed within.

Figure 4.10 Opening screen. *Phone Story* by Molleindustria (Molleinustria, 2011).

Screen 2:

Once upon a time... there are minerals resting in the bowels of the earth. One of these minerals, called Coltan, is found in most electronic devices. The majority of coltan's world supply is located in the Democratic Republic of the Congo ... The increasing demand for Coltan produced a wave of violence and massacres in the Congo. Military groups enslaved prisoners of war, often children, to mine the precious material. Directly or not, we are all involved in this complex illegal activity.

Figure 4.11 Level 1: Children mining for coltan. *Phone Story* by Molleindustria (Molleinustria, 2011).

Level One "forces" me to take on the role of military oppressor and overlord directing what seems like Congolese children to dig for precious materials. To achieve my goal of passing the level, I must make the soldiers threaten the children with a rifle and yell out orders to work harder (assumedly) in a "Wha, Wha, Wha" character voice. The action is at once both comedic and horrifying. My actions as a player, passing my fingers up and down the glass screen, implicate me in the very actions that allowed the device to be made. By owning this device, I am remade into that oppressor, pushing on these helpless children in their inhuman bondage.

Screen 3:

This phone was assembled in China inside a factory as big as a city. The people working there are constantly subjected to abuse and discrimination. They work in inhumane conditions ... Over the span of a few months, more than twenty workers committed suicide out of extreme desperation. We addressed this issue by installing suicide prevention nets.

Figure 4.12 Level 2: Factory worker suicides. *Phone Story* by Molleindustria (Molleinustria, 2011).

Level Two moves me to another part of the world and into another ethical dilemma: Help save the lives of those oppressed workers who assemble the phone in monolithic factories like that of Foxconn in China. To save their lives, I must push a stretcher back and forth across the screen to catch the falling workers who have plummeted by choice to their death, out of sheer desperation at their working conditions in the factory. The gameplay here is much like playing the old Atari game *Pong*, except when I miss a falling factory worker, they fall flat on the ground with a deafening bloody splat, resulting in a deduction of my goal meter. The difficulty in controlling the stretcher makes me momentarily forget that the purpose of catching these exploited workers is to simply exploit them further. Upon catching, they are sent back into the factory. I assume they are put back to work. Again, like Level One, I am engaged in the flow of play with the attempt to achieve my goal of beating the level, and as soon as I do, I immediately recognize again how the game is using the operative logic of the device to highlight my role in the ethical problematics of the device's production.

Screen 4:

You purchased this phone. It was new and sexy. You've waited for it for months. No evidence of its troubled past was visible. Did you really need it? Of course you did. We invested a lot of money to instill this desire in you. You were looking for something that could signal your status, your dynamic lifestyle, your unique personality. Just like everyone else.

Figure 4.13 Level 3: Consumers on new release day. *Phone Story* by Molleindustria (Molleinustria, 2011).

Level Three is a bit closer to home, figuratively and geographically. The mini game has me take on the role of the stereotypical Apple Store employee, dressed in a T-shirt and jeans with the iconic white lanyard around my neck. It is release day for the latest iteration of smartphone, and my task is to pass out the device to the throngs of deranged technophiles who have been waiting outside the store for days to be the first to get one. I must literally throw the devices at them so that they do not run face first into the store's façade in their consumerist delirium. Other than the difficulty of the actual game mechanics, this level seems more comic than critical in design, but its narrative picks apart the capitalist logic of techno-culture, shredding my own humanity to the bone. I am just like virtually every other iPhone owner; I upgrade nearly every cycle out of sheer narcissism and fear of becoming behind the trends.

Screen 5:

Soon we will introduce a new model that will make this one look antiquated and you will discard it. They say they will recycle it but it will probably be shipped abroad to places like Ghana, Pakistan or back to China. There its materials will be salvaged using methods that are harmful to human health and the environment. Parts of this phone will contaminate air and water. Others will reincarnate into new products.

Figure 4.14 Level 4: Recycling the First World's trash. *Phone Story* by Molleindustria (Molleinustria, 2011).

Level Four returns me to one of the "forgotten" regions and invisible peoples of the world. An endless conveyor belt of scrap metal, computer parts, and discarded "obsolete" iDevices moves down the screen, where depressed people of color wait on the sides operating rustic ovens and cauldrons. Using the swipe of my finger, my task is to pass off the materials to these avatars at a feverish pace so that they do not simply fall into the abyss at the end of the belt. The level makes visible the arduous and non-gratifying task it must be to repurpose the trash of the First World. And what am I rewarded with when completing the level? A screen that simply lets me know that the cycle is not complete. I now enter "Obsolescence Mode."

Screen 6: *And the cycle continues.*

The next ten levels (and I assume all levels until infinity) simply repeat the first four with a new animation at the beginning. Each animation shows a shadowed smartphone-shaped device descending from the heavens in black outline to the earth, moving to a digitalized rendition of the opening music from *2001: A Space Odyssey.* Each successive level amps up the irony and the critique of the smartphone as a not-so-unique agent in the consumerist network of planned obsolescence. I am made uniquely aware at the beginning of each level that I am part of that network, whether I like it or not.

Level 1: iThing Beta
Level 2: iThing 2.0
Level 3: iThing 3g
Level 4: iThing Max
Level 5: iThing Special Edition
Level 6: iThing ∞
Level 8: iThing +
Level 9: iThing π
Level 10: iThing Air
Level ....

Conclusion: The iDevice, the Consumer, and Meaningful Play

One of the complications of the technogenetic paradigm of iDevices is the liminal positions set up for the mediatized spectator. When we enter into the liminal position that is play, "we are compelled into subjective positions that both are and are not our own; we are entering worlds, more precisely, spaces, that both are and are not our own" (Boulter 2015, 23). Games and game design structures are one way that experiential spectatorship can and does adapt to this paradigm. This is where the importance of games as cultural determinates of social conditions becomes important to take seriously. Mary Flanagan (2013) explains:

> Games are artifacts of historic and cultural importance, but they are also something *beyond artifact* in that games also function as a set of activities that carry conventions like audience role, interaction, currency, and exchange. They are systematic causal correspondences between particular design features in games that indicate specific social conceptualizations and outcomes. (259, italics in original)

As an architecture developed under the technic of pervasive connectivity, *Game Play* frames and dictates the operation of many experiential performances, replicating the interaction between mediatized spectators and iDevice technology. One of the hallmarks of twenty-first-century games is how they blend spectatorship with mediated and technological objects through mediated and cultural participation (151). The assemblage of cultural, technological, and performative elements of gamified spectatorship resembles a social configuration of mediatized subjectivity. Performances structured around meaningful and critical game play offer spectators the possibility of ludic critical exchange correlating with their everyday condition in mediatized life. In cultures led by iDevice-based media streams, a mediatized spectator becomes a fusion of the media object (game, performance), the technological tool (phone, tablet, etc.), and their perceptual apparatus made up of the body/brain/environment assemblage. This fusion is part of a mediatized paradigm, but the message or focus of the game played is what allows for a meaningful experience.

Each of the examples above meaningfully uses ludic criticality as a tool to make visible the player's place with(in) their mediatized social situation. Visibility allows both the tool and the player to perform acts of social reflection and possibly social change. *Pokémon GO* does this by covertly hiding histories and cultural accounts beneath the digital landmarks used to gamify the world around us. *Adventure One* similarly uses the iDevice to guide one toward a gamified objective only found by straddling the line between fact and fiction. *Phone Story* is unique, however. By critiquing its own logic and place in consumer culture, it gamifies the game itself. The graphic dark humor used as part of the game's critique also caused it to be pulled from Apple's App Store days after it was released. I had to purchase a separate Android-based phone just to play the game for this book. My ethical position in conducting the very research implicated myself further into a commercial and corporatized system in which the game would use play tactics to critically evaluate. Apple cited that the game violated some of its app guidelines as justification for the banning (Brown 2011). By pulling the game from the store, Apple asserted its authority to control the narrative about its most popular product. The removal also increased the critical subversive power of the game's narrative, allowing its players—assuming they could access the game via Android and the Google Play store—a heightened awareness of the game's message. The game hijacks the smartphone as a subversive and performative tool, allowing the player to engage in a form of mediatized ethics by better understanding how the device itself intricately connects them to all the links in the supply and production chain. Apple's refusal to publish the game amplifies the ludic critical exchange found in the experience of the game by giving credence to the game's message.

The message delivered to the player when failing to complete a level fully sums up the player's ethical position and relationship with both the game, narrative, and technology. Simply, as an owner and user of the device, you are complicit. The mediatized spectator is complicit in the action of the performance, and when they address ludic criticality through the performance's operation, the player has the ability to re-address how it will navigate the future in a manner that may or may not fulfill ethical responsibilities in a truly interconnected world. I agree with Flanagan (2013) when she argues that "If digital artifacts have truly become a magic circle in which players enter a sanctioned play space, then this culture of play, or play culture, as it is commonly termed, is one in which participants find a space for permission, experimentation, and subversion" (13). The hybrid liminal space opened up by the iDevice allows ludic criticality to occur as a perpetual engine of self and world examination via experiential spectatorship.

Game Over Screen:

You didn't meet the goal.
Don't pretend you are not complicit.

Figure 4.15 End of game. *Phone Story* by Molleindustria (Molleinustria, 2011).

Notes

1 King et al. (2013) apply "nomophobia" to the social anxiety condition associated with disconnecting from one's access to virtual space and time.
2 Socioeconomic status plays a large role in the use of mobile device as primary conduit to the internet.
3 The term *iGeneration* was first introduced to mass audiences through the song "iGeneration" by MC Lars in 2006 (https://vimeo.com/19497936)
4 Dedicated e-readers stopped being tracked in 2016.
5 The first mobile communications devices that resemble the cell phone were introduced by the United States military in the 1930's. These devices were similar to one-way paging devices or "walkie talkies." The first mobile phone service was introduced in 1947 by Bell Systems.
6 E-readers are not covered here because they are largely used to download material for "off-line" consumption.
7 The *Book It* program still exists. For more information see: https://www.bookit-program.com/about
8 For a step by step experiential walk through of a player's experience of one of the games stages see Elizabeth Tobey's (2016) blog post at: https://medium.com/@dahanese/a-game-called-the-jejune-institute-d12cc17a85d0

9 Three teams are in constant battle for the colonization of geographical locations.
10 An abbreviated version of this Instance first occurred in (Lewis 2017).
11 For an example of the game's structure and gameplay see: https://www.youtube.
com/watch?v=sSMSFLAsNzc. (Koz and Phonestory.org 2011).

Bibliography

Abt, Clark C. 1970. *Serious Games*. Viking Press.

Alston, Adam. 2016. *Beyond Immersive Theatre: Aesthetics, Politics and Productive Participation*. Palgrave.

Arnold, Ross D, and Jon P. Wade. 2015. "Definition of Systems Thinking: A Systems Approach." *Procedia Computer Science* 44: 669–78.

Baudrillard, Jean. 2014. *Screened Out*. Translated by Chris Turner. Verso.

Boulter, Johnathan. 2015. *Parable of the Posthuman: Digital Realities, Gaming, and the Player Experience*. Wayne State University Press.

Bourriaud, Nicolas. 2002. *Relational Aesthetics*. Translated by Simon Pleasance, Fronza Woods, and Mathieu Copeland. Les presses du reel.

Brandon Hunter, Lindsay. 2021. *Playing Real: Mimesis, Media and Mischief*. Northwestern University Press.

Brown, Mark. 2011. "Apple Bans Phone Story Game That Exposes Seedy Side of Smartphone Creation | WIRED." *Wired.com*. 2011. https://www.wired.com/2011/09/phone-story/

Caillois, Roger. 1963. *Man, Play, and Games*. Thames and Hudson.

Calleja, Gordon. 2011. *In-Game: From Immersion to Incorporation*. MIT Press.

Coney represented by, Will Drew, and Stevens Tassos. 2015. *Adventure One*.

Coney. n.d. "Coney." http://coneyhq.org/about-us/

Costikyan, Greg. 2002. "I Have No Words & amp; I Must Design: Toward a Critical Vocabulary for Games." *Frans Mäyrä. Tampere.* http://www.costik.com/nowords2002.pdf

Couldry, Nick, and Andres Hepp. 2017. *Mediatized Construction of Reality*. Polity.

Deterding, Sebastian. 2012. "Gamification." *Interactions* 19, no 4: 14–17.

Díaz, Lily. 2019. "From Simple Rules to Complex Performances—Interview With Blast Theory'S Matt Adams," 240–248. In *Ubiquitous Computing, Complexity, and Culture*. Routledge.

Ekman, Ulrik, Jay David Bolter, Lily Díaz, Morten Søndergaard, and Maria Engberg. 2016. *Ubiquitious Computing, Complexity and Culture*. Routledge.

Farman, Jason. 2010. "Hypermediating the Game Interface: The Alienation Effect in Violent Videogames and the Problem of Serious Play." *Communication Quarterly* 58, no 1: 96–109.

Farman, Jason. 2012. *Mobile Interface Theory: Embodied Space and Locative Media*. Routledge.

Farman, Jason. 2014. *The Mobile Story: Narrative Practices with Locative Technologies*. Routledge.

Farman, Jason. 2015. "Stories, Spaces, and Bodies: The Production of Embodied Space Through Mobile Media Storytelling." *Communication Research and Practice* 1, no 2: 101–16.

Farman, Jason. 2021. *Mobile Interface Theory: Embodied Space and Locative Media*. 2nd ed. Routledge.

Firth, Jordan. 2015. *Smartphones as Locative Media*. Polity.

Flanagan, Mary. 2013. *Critical Play: Radical Game Design*. MIT Press.

Flynn, Jack. 2023. "20 Vital Smartphone Usage Statistics [2023]: Facts, Data, and Trends on Mobile Use in The U.S. - Zippia." *Zippia.com*

Freshwater, Helen. 2009. *Theatre and Audience*. Palgrave.

Greengard, Samuel. 2015. *The Internet of Things*. MIT Press.

Handley, Lucy. 2019. "Smartphones: 72% of People Will Use Only Mobile for Internet by 2025." *CNBC.com*. January 24, 2019. https://www.cnbc.com/2019/01/24/smartphones-72percent-of-people-will-use-only-mobile-for-internet-by-2025.html

Hayles, N. Katherine. 2012. *How We Think: Digital Media and Contemporary Technogenesis*. University of Chicago Press.

Henderson, Clint. 2017. "2017 Mobile Marketing Statistics (Trends, Predictions, & Mobile Strategy) - Wired SEO." *Wiredseo.com*. 2017. https://www.wiredseo.com/mobile-marketing-statistics-2017/#5

Herbrechter, Stefan. 2013. *Posthumanism: A Critical Analysis*. Bloomsbury.

Hjorth, Larissa, and Ingrid Richardson. 2014. *Gaming in Social, Locative and Mobile Media*. Palgrave.

Hjorth, Larissa, and Ingrid Richardson. 2020. *Ambient Play*. MIT Press.

Huizinga, Johan. 1955. *Homo Ludens: A Study of the Play Element in Culture*. The Beacon Press.

IDC Custom Solutions. 2013. "Always Connected for Facebook Case Study." Idc. Com. 2013. https://www.idc.com/prodserv/custom_solutions/download/case_studies/PLAN-BB_Always_Connected_for_Facebook.pdf

Kelly, Kevin. 2016. *The Inevitable: Understanding the 12 Technological Forces That Will Shape Our Future*. Viking Press.

King, A.L.S, A.M. Valenca, A.C.O. Silva, T. Baczynski, M.R. Carvalho, and A.E. Nardi. 2013. "Nomophobia: Dependency on Virtual Environments or Social Phobia?" *Computers in Human Behavior* 29, no 1: 140–44.

Koz and Phonestory.org. 2011. "Phone Story: An iPhone Educational Game. Now Banned from the AppStore" *Youtube.Com*. https://www.youtube.com/watch?v=sSMSFLAsNzc.

Lewis, William. W. 2017. "Performing 'Posthuman' Spectatorship: Digital Proximity and Variable Agencies." *Performance Research* 22, no. 3: 7–14.

Lewis, William W, and Sarah Johnson. 2017. "Theatrical Reception and the Formation of Twenty-First-Century Perception: A Case Study for the iGeneration." *Theatre Topics* 27, no 2: 123–36.

Machon, Josephine. 2013. *Immersive Theatres: Intimacy and Immediacy in Contemporary Performance*. Palgrave.

Martin, Adam, Brooke Thompson, and Tom Chatfield. 2006. "Igda-Alternaterealitygames-Whitepaper-2006.pdf."

McGonigal, Jane. 2011. *Reality Is Broken: Why Games Make Us Better and How They Can Change the World*. Penguin Books.

McNish, Jacquie, and Sean Silcoff. 2015. *Losing the Signal: The Untold Story Behind the Extraordinary Riser and Spectacular Fall of Blackberry*. Flat Iron Books.

Merleau-Ponty, Maurice. 2013[1948]. *Phenomenology of Perception*. Translated by Donald Landes. Routledge.

Molleinustria. 2011. *Phone Story*. *https://www.phonestory.org*

Perrin, Andrew. 2016. "Book Reading 2016." *Pewinternet.org*. http://www.pewinternet.org/2016/09/01/book-reading-2016/

Perrin, Andrew, and Sara Atske. 2021. "About Three-in-Ten U.S. Adults Say They Are 'almost Constantly' Online." *Pewresearch.org*. https://pewrsr.ch/2Y5pwdX

Pew Research Center. 2017. "Mobile Fact Sheet." *Pewresearch.org*

Pew Research Center. 2021. "Mobile Fact Sheet." *Pewresearch.org*. https://www.pewresearch.org/internet/fact-sheet/mobile/

Poláková, Petra, and Blanka Klímová. 2019. "Mobile Technology and Generation Z in the English Language Classroom: A Preliminary Study." *Education Sciences* 9, no 3: 203.

Rosen, Larry. 2010. *Rewired: Understanding the iGeneration and the Way They Learn*. Palgrave.

Salen, Katie, and Eric Zimmerman. 2004. *Rules of Play: Game Design Fundamentals.* MIT Press.

Schechner, Richard, and Sara Brady. 2017. *Performance Studies: An Introduction.* Third edition. Routledge.

Souza e Silva, Adriana de. 2009. "Hybrid Reality and Location-Based Gaming: Redefining Mobility and Game Spaces in Urban Environments." *Simulation and Gaming* 40, no 3: 404–24.

Suits, Bernard. 1990. *Grasshopper: Games, Life, and Utopia.* David R. Godine.

Sutton-Smith, Brian and Elliot Avedon. 1971. *The Study of Games.* J. Wiley.

Tobey, Elizabeth. 2016. "A Game Called the Jejune Institute." *Medium.Com.* February 26. https://medium.com/@dahanese/a-game-called-the-jejune-institute-d12cc17a85d0

Twenge, Jean M. 2017. *iGen: Why Today's Super-Connected Kids Are Growing Up Less Rebellious, More Tolerant, Less Happy — and Completely Unprepared for Adulthood*/*and What That Means for the Rest of Us.* ATRIA Books.

Twenge, Jean M. 2023. *Generations: The Real Differences Bewteween Gen Z, Millenials, Gen X, Boomers, and Silents – and What They Mean for America's Future.* Atria Books.

White, Gareth. 2016. "Theatre in the 'Forest of Things and Signs.'" *Journal of Contemporary Drama in English* 4, no. 1: 21–33.

Wijman, Tom. 2024. "Newzoo's Games Market Revenue Estimates and Forecasts by Region and Segment for 2023." *Newzoo.com.* February 8, 2024. https://newzoo.com/resources/blog/games-market-estimates-and-forecasts-2023#:~:text=The%20global%20games%20market%20reached,begin%20to%20grow%20once%20again

Wilken, Rowan. 2014. "Proximity and Alienation: Narratives of City, Self, and Other in the Locative Games of Blast Theory." 175–91. In *The Mobile Story: Narrative Practices with Locative Technologies*, edited by Jason Farman. Routledge.

Wilken, Rowan, and Gerard Goggin. 2015a. *Locative Media.* Routledge.

Wilken, Rowan, and Gerard Goggin. 2015b. "Locative Media: Definitions, Histories, Theories." 1–19. In *Locative Media.* Palgrave.

Zimmerman, Eric. 2015. "Manifesto for a Ludic Century." 19–22. In *The Gameful World: Approaches, Issues, Applications*, edited by Steffen P. Walz, and Sebastian Deterding. MIT Press.

5 The Architecture of Role Play

5.0 Algorithms, Avatars, and Affective Computing: Datafication and Social Feedback Loops

How well do you think you know yourself? How well do you think others know you? Is your personal understanding of your own sense of self simply a reflection of what these others think? What does it mean when these others are not people but instead machines, corporations, or government entities? How do your daily experiences with these various others help to shape the way you perform your own sense of self? Is it possible that there are multiple versions of you performing in their own multiplied constructions of reality? If so, how do you know which one of these roles is the true you? Is it possible that in the face of digitalization there is no such thing as a true and undivided self? Is this self simply an amalgamation of unique data points assembled through the variety of online and offline interactions you encounter daily? If so, how can you ever truly define yourself as anything other than data?

This chapter focuses on the technic of *data determinism* and its relationship to the *Architecture of Role Play* which works as a framing system for performances of selfhood both in everyday reality and in constructions of experiential spectatorship. Data determinism is brought forth through algorithmic systems serving as the invisible underlayer of most of our current mediatized realities. As the final architecture of exchange included in this model for experiential spectatorship, role play operates as both a relational and performative way of life and subsequently a mode of spectatorial exchange staged when peoples' daily interactions become entangled with(in) algorithmic structures. These interactions offer mediatized spectators the constant potential to develop multiple notions of selfhood while at the same time threatening to limit this potential. Role play arises in the interactions between data-based technologies and mediatized spectators by actively shaping their social reality as a form of techno-performativity with(in) paradigms of datafication. This process for structuring social reality serves as the most recent wave of mediatization. It is also embedded inside digitalization, leading to the full force of deep mediatization. This wave is based on an understanding "that 'data' and 'information' generated by systems of computers

DOI: 10.4324/9781003398813-6

are today a precondition for everyday life" (Couldry and Hepp 2017, 123). Through datafication, all media become subsumed and reconstituted through processes of data collection and data analysis, often described as dataveillance, a form of data-based tracking and profiling (Esposti 2014; Harding 2017; Morrison 2016). As quantifiable points of information and identification, data becomes the meta-medium for all mediatized processes of life and operates as the basic element for structuring one's sense of selfhood. This selfhood is constructed through what John Cheney-Lippold (2017) describes as an "interpolative method" where subjecthood and thus selfhood,

> is not identified according to a preexisting, one-to-one relationship of one's self. Rather, that subject is understood in the context of statistical estimation that fills in the holes of existing data with new algorithmic approximation ... Ultimately, this subject can be expressed as a type of composite algorithmic identity: the particularity of one's individual identity is replaced by an aboutness of one's identifications. (170)

The algorithmically generated self is therefore an identity performing with(in) datafication. It performs its constitution through data flows. Following the arguments of performativity (Butler 1988, 2010) augmented through technologies, this selfhood is therefore a construct that is always fluid and changing based on one's societal influences in relation to their data-based interactions. As such, selfhood under datafication is therefore constantly in relation to data through the technic data determinism and the technological processes contained within.

Data determinism works as a final intersubjective micro layer of deep mediatization affecting our current era and, as such, relies on each of the previous technics discussed. I introduce data determinism to explain how role play serves as a form of experiential spectatorship through which spectators may gain opportunities to construct multiple mediatized selves with the capacity to work against data-based systems. When role play is creatively harnessed as spectatorship in performance, it allows a rehearsal of these selves that can be actualized beyond a performance frame. For example, by imaginatively or physically role playing as a different version of one's "true self" within a video game, theatrical event, or interactive narrative, the spectator is offered an understanding of the flexibility of one's personal self as a mode of performance within flux. By doing so, they may then rethink the way they deploy this understanding when interacting with algorithmic systems in the actual world of their everyday life. As argued in the previous chapter, this life has already become gamified through their connection to technologies of pervasive connectivity, but now role play offers a new performative tool through which to "game the system." When utilized as a lens for audience research, correlating spectatorial acts to data determinism can also assist in better defining the experiential aspects of role play in with various experiential media.

Before discussing the architecture and its foundations, I cover the socio-technical paradigm that allows data determinism to become a prevailing technic of the early twenty-first century. Like the previous technics, data determinism has a social shaping impact through cognitive effects on the mediatized spectator's perceptual apparatus. What is different is how systems of data operate invisibly in the background which complicates the formal operations of relationality. Where the other frames for mediatized spectators relied on a constant conscious relation to their technologies and technics, data determinism disrupts the circuit of inputs and outputs to produce a system where the human side of the relation becomes unaware of their part in the infrastructure of data. The invisibility of the infrastructure is a fundamental basis of deep mediatization as it fades into the background and becomes impossible to separate from. The process is also multilayered which warrants quoting Couldry and Hepp (2017) in detail. They state:

> Infrastructures are, at their root, tools for human action operating at the highest level of complexity, a black-boxed substrate of "ordinary" human action. In the digital world, our infrastructural tools (for example, our pages on a social media platform) are increasingly entangled in *powerful* and *distant* processes, which we cannot unpick or challenge … many of today's "digital tools", as we use them, are black-boxes of a different sort, black-boxes that *are also in the act of using us*. They track our actions algorithmically, not to enable the tool to work better for us, but to generate data for the *toolmaker's* use … Today's data-based artefacts now themselves operate to *typify* humans mostly for commercial ends and surveillance, to construct a seamless world for commerce and control. We might refer to this as *tool reversibility* … Whenever we use a data-based tool, it is already using us. (131–132, emphasis in original)

When they refer to "using us," they mean that algorithmic systems generate new black-boxed—opaque and with an operational construction with an unknowable basis—realities that are only comprehensible by the machines that create them. These are realities founded on the logic of the database, where human cognition operates too slowly to keep up with the tools reading and shaping these human realities. To best understand this process of being used by our algorithmic tools, I will put into conversation Mark B. N. Hansen's (2015) process of worldly sensibility and N. Katherine Hayles' (2017) description of the cognitive nonconscious which are constantly in relation to and hence shaped through the power of data determinism.

Once this foundation is laid, a brief foray into the many digital tools that aid in the construction of our data double is explored. Of particular interest is the relationship between data determinism and the Internet of Thing (IoT). A data double is the digital equivalent of machine-readable instances of human selfhood which bring about constant unconscious micro-acts of creative human self-formation. Each of us has a multitude of double based on the various data interactions we have. The data double has been described as a "data-body" (Lucie 2021), a "dividual" (Deleuze 1992), and a "data derivative" (Matzner 2019). In each case, there is a data-based construction created by algorithmic surveilling technologies to develop forward predicting behaviors that governing one's future actions through a form of Foucauldian biopolitics and biopower. Unlike the tools and technologies of the previous chapters, algorithmic tools are largely self-regulating and are thus processual and not as easily "graspable" like a physical smartphone or browser interface that allows one to upload personal content. Instead, they are tools for hidden algorithmic processing. These tools might have a front facing material construction—for instance, an Alexa enabled smart speaker—but their operations are largely invisible as they perform their tasks covertly through tactics of surveillance. By categorizing and explaining what these tools are and how they operate in the background of daily life, we can gain an understanding of the materialist phenomenology required to analyze role play as acts of spectatorial performativity under paradigms of data determinism. The closing part of this section moves to the *Architecture of Role Play* which allows mediatized spectators a form of *ludic creative agency* and *exchange* through which they might engage in the purposeful formation of playful selves in performance. Through ludic creative exchange, these spectators act as (co) authors of their own realities, which presents a set of tools for rewiring the circuit of relations with algorithmic systems.

Non-Conscious Cognition, Worldly Sensibility, and Data Determinism

While progressing through this project, I have highlighted communication technologies and media that have gradually become more embedded and integrative within social constructions of mediatized reality. These technologies have allowed the formation of technics which then enact fundamental evolutions in the perceptual apparatus of mediatized spectators. Each of the previous technics has operated in a historical trajectory of overlap as building blocks, working off each other in certain capacities. The focus of this chapter is on technologies that support the technic of data determinism, a propelling force that is brought about through aspects of algorithmic surveillance, filtering, and feed-forward, where the construction of a stable unified self is put into constant question.

Following Hansen (2006, 2015, 2018) and Hayles (2014, 2017), technological media based on algorithmic constitution—whether they are employed through aspects of social media, voice activated digital assistants, governmental and corporate profiling systems, GPS tracking software, and early-stage artificial intelligence platforms—operate at a near invisible level that is beyond the reach of conscious human sensing. Algorithmic systems, like those embedded in the platforms above, have the power to invisibly control the formation of one's sense of self by hiding the fact that they even exist. As an underlying structure that has helped support the previous technic of pervasive connectivity, these algorithmic systems operate in the background, gaining their stealth capabilities and grow more powerful as they become ubiquitous and common. Being invisible does not negate their effective and affective impact, however. In fact, it makes their ability to subjectivize human beings even stronger. The techno-social paradigms of digitalization and datafication cause social systems to develop in constant reflexive relationality with media technologies, where "the production of the self within [these] digital cultures relies on a self-illusion, which obscures its technological operations, while at the same time binding the human to them" (Leeker 2017, 33). By interacting with algorithmic systems, one's relation to the world becomes entangled with the subjectivizing power of the algorithms' programmed intentionality.

Algorithmic entanglement becomes common place with digital assistants like Siri and Alexa, that are fundamentally anthropomorphized search engines algorithmically programmed to deliver personalized results. New constructions of selfhood are manipulated through data determinism, where the algorithms that run these digital assistants attempt to learn our behaviors and categorize our future potentials. To operate best, they must be trusted as a tool that wishes to assist, to help, to make our lives better. They do this by hiding the fact that they are first a surveilling tool geared toward creating data doubles based on the input of their users. By becoming a "friend," they appear to be something other than a collection tool, and this allows them to operate indiscriminately. Media theorist Tobias Matzner (2019) explains what algorithms embedded in these tools are really doing:

> They take important decisions, promise novel insights into huge troves of data, distribute goods and services, classify persons (potential partner, customer, criminal), try to detect terrorists and much more. A lot of this is done automatically, reacting to input in a "smart" or "intelligent" way. Thus, algorithms take positions or functions that used to require humans—or even have been impossible as long as humans were the only intelligent actors. (123)

This quote highlights the agential potential of algorithms and the power they perform within the intricate network of interactions with humans and other non-human objects by exploring and complicating the boundaries between users and algorithms to destabilize the possibility of a boundary under datafication. Matzner explains, "I think there is no clear-cut boundary between humans and algorithms [...] this boundary is structured by a productive tension of continuity and difference" (127). To fully comprehend the power of algorithmic systems, one must expand their own understanding of relationality that occurs in the interactions between the individual and the machine.

By deconstructing the boundary between mediatized spectators and algorithmic systems, one can better understand the influence of data determinism. Hansen (2015) describes how twenty-first century media—of which algorithms are a prime example—enact a "fundamental transformation caused by the complex entanglement of humans within networks of media technologies that operate predominately, if not almost entirely, outside the scope of human modes of awareness (consciousness, attention, sense perception, etc.)" (5). He builds his argument through the process philosophy of Albert Whitehead (1978), the basis of which reassembles the prevailing notion of subjectivity as centered in notions of human agency and human consciousness. For Hansen, this subjectivity expands agency into action that "implicates *the entirety of the* universe, whether through actual engagement (via 'positive prehension') or through active exclusion (via 'negative prehension,' which, crucially, is still a form of exclusion that is still a relation)" (Hansen 2015, 10). Prehension is Whitehead's chosen terminology for feeling or affect in the logic of sensing something that can be attuned or grasped through objective means. You find this root in compound words such as comprehension (bringing together sensing/to understand with or to grasp a sense) and apprehension (sensing ahead/sensing ahead of understanding). A "positive prehension" requires the acceptance of the sense (to grasp something and hold onto it), while a "negative prehension" is a disavowal of that sense (to sense but reject or ungrasp). In either case, there is a notion that there is something there: The difference is whether that sensing is held onto and grasped—as in consciously perceived—or, that sensing is rejected but still existing within the sub-conscious functions of the individual without passing through the filter of consciousness as such. Both aspects of sensing have agency. It is important to also note that prehension is not exclusively a process imbued through human consciousness. Non-human objects such as algorithms can prehend as well, and in today's data environment, they can do so more effectively.

Hansen defines a worldly sensibility built off a breaking down of the divide between subjectivity and objectivity into simply agential relationality. In this sensible world, there is no divide between subject and object, but rather, all instances are objects that gain subjectivity through the relational operation with each other. In this way, there is no original subject or even original object to compare against, and therefore, an infinite number of possibilities and potentialities which we call subjectivities that arise out of the relational process between quantifiable data points. To an algorithm, people are simply objects to subjectify based on processes of computational comparison and correlation, which propel data relationality and subsequently techno-performativity. Algorithmic comparison and correlation act as a form of techno-performativity when linked to the self-forming properties of human consciousness. In the conventional understanding of human performativity, explored extensively in performance studies, there is a link between "intentionality, reflexivity and sense-making, to embodiment, repetition and transgression" (Leeker, Schipper and Beyes 2017, 11). Technological performativity "on the other hand, refers to deterministic operation without semiotic or affective qualities" (Leeker, Schipper and Beyes 2017, 11). Algorithmic systems enact this form of performativity through their processes of sensing human inputs and developing "datafied subjects" of their own (Cheney-Lippold 2017). In techno-performativity, all humans act as objects standing blank, waiting to become subjects through a process of performing with all other technological objects. In algorithmic social worlds, identity as data is one of these objects. This chapter intends to show that mediatized spectators can still affect their construction of personal selves through these processes within techno-performativity.

The overlapping of both human and algorithmic realities allows for a mediatized subjectivity to arise out of the multiple assemblages of agency and relationality. Hansen (2015) states, "Our distinctly human subjectivity is the result of a complex assemblage of overlapping, scale-variant microsubjectivities functioning distinctly autonomously" (11). From this perspective, all experiencing becomes a neutral function due to the multiplicity of subjectivities available to both human and non-human objects and the constant interplay between these multiple subjectivities. Each subjectivity has a unique capacity for creating multiple experiences to order a reality from and each can be enacted through the Architecture of Role Play. Before the potential of a data-based technological construction of performativity existed, the prevailing understanding of subjectivity—one that arises from conscious perceptibility of experience—was Cartesian, with a human being at the center of the subjectivizing process, or what Hansen refers to as the "human as 'experiencer par excellence'" (14, quotations in original). Using the intersubjective framework of mediatization allows one to examine acts of experiencing as a relational and non-anthropocentric process, where expanded configurations between agency, power, and the formations of selfhood implicate algorithmic systems which are non-material objects. The mode of mediatized subjectivity Hansen argues for is precisely the theoretical model necessary when considering the assemblage of machine, data, and human that occurs under processes of data determinism. Under this model, the project of understanding one's place in the world is to "complexify the human by multiplying its connections, not to wall it off as a helplessly imperializing intentionality" (17). As people become aware that they are "implicated in every element of the universe" (16), a data determined subjectivity emerges and offers the potential for multiple constructions of selfhood found via **purposeful acts of role play**. These multiples are based on data but can be harnessed by both machines and their human users. Unfortunately, accessing these new subjectivities is increasingly difficult because they exist outside the realm of current human perception and consciousness due to data determinism. This is due to what Hansen calls feedforward, which is a crucially important process to understand when discussing the tension between people and algorithms. I've discussed how feedback is a crucial part of experiential spectatorship and mediatized social life.

Feedforward works as "the operation through which technically accessed data of sensibility enters into futural moments of consciousness as radical intrusions from the outside" (30). Put more simply, feedforward is the predetermination of conscious awareness delivered by machine cognition. In feedforward, intelligent machines gain the power to calculate future human response through interactions with user-generated data.

Hansen also argues the ontological basis of twenty-first century media, which machine reading algorithms are a part of, is different than those of the twentieth century in its "shift from a past-directed recording platform to a data-driven anticipation of the future" (4). Twenty-first century media does more than record and respond to the world; it calculates, predicts, and reifies instances of reality. Hansen explains the dangers of the current media environment:

> It has become markedly less benign over the past decade as Google has consolidated its monopoly over Internet searching and data aggregation, the process of perfecting a system for extracting data-value form our every web search; as Facebook has consolidated its monopoly over sociality on the Internet, in the process perfecting a system for extracting consumer profiles ripe for delivery to advertisers; and in general, as today's media industries have honed methods for mining data about our behavior that feature as their key element the complete bypassing of consciousness... (4)

Consciousness is currently unique to some biological objects such as human beings but is only a secondary process connected to cognition which now must also include technological objects due to data determinism.

Human consciousness arises and is constituted out of "the interaction of multiple components of the human with the world, even as the world is produced for the human due to this consciousness" (Nayar 2014, 41). This conceptualization of consciousness is at the heart of the phenomenological operation of a mediatized spectator's perceptual apparatus. As explained before, the components of the perceptual apparatus are the brain and body—as interconnected and dual-directional sensing technologies—as well as the non-biological techno-bodies in the environment sensed by the human part of the perceptual apparatus. What makes human consciousness such a complicating factor in the subjectifying process of reality creation is how the world is only made visible to humans because of our inherently subjective process of interacting (sensing and responding) with it. Consciousness forms from both sensing and creation of the environment we are a part of. Cognition, on the other hand, senses and determines the variables that lead to human "creation" of the world prior to consciousness. It is the underlying biologically encoded algorithm of human perception. Human beings cannot "exist" as unique individual selves without consciousness, and the social worlds humans experience consciously only exist because of consciousness' capacity to filter out and create distinct realities. The environment and the variables that allow the formation of both are constantly available to all sensing objects and are in flux. Consciousness is then both the technology and process that filters out most data made perceptible to humans through cognition. The filtering process "creates the (sometime fictitious) narratives that make sense of our lives and support basic assumptions about worldly coherence" (Hayles 2017, 9). Consciousness, is therefore, a technology that makes our reality coherent in any particular instance. For example, when referring to VR sickness in an earlier chapter, the reality experienced by the immersant's perceptual apparatus became incoherent because visual stimuli (data) did not match up with corporeal response, causing a rupture in the perceptual apparatus' sensing function. When all levels of the perceptual apparatus work in sync "properly," the world experienced makes sense. It becomes comprehensible.

Consciousness and unconsciousness are then second-order procedures existing primarily as technologies of making-sense of the world, just as autopoiesis and subsequently performativity are second-order procedures of making-sense of the self in the world. In this way, consciousness is similar to algorithms when applied to media that performs with this certain form of specificity. Consciousness is only one part of the larger operation of cognition, which Hayles (2017) assigns to all objects within Hansen's world of sensibility. Hayles defines cognition as "*a process that interprets information within contexts that connect it with meaning*" (22, italics in original). Consciousness is therefore the process where meaning becomes realized for the human post cognition. Consciousness is a subordinate worlding process because it operates after—both procedurally and temporally—cognition, remaining always in the past of actual sensing. Cognition is a first-order operation engaged in both the collection of sense data and also the processing of that data for extrapolation out into both conscious and unconscious systems of world creation. Hayles uses the term nonconscious cognition when discussing the ability of both biological and technical objects to engage in this first-order operation. She states that the process of "nonconscious cognition operates at a level of neuronal processing inaccessible to the modes of awareness but nevertheless performing functions essential to consciousness" (10). Because of its first-orderliness, nonconscious cognition is "a mode of interacting with the world enmeshed in the 'eternal present'" (1, quotations in original). Nonconscious cognition exists in the *now of being* vs consciousness' *to be of being*.

Hayles also explains how machine cognition and human world-sensing and interpretation interact using the term nonconscious cognition:

> Nonconscious cognitions are increasingly embedded in complex systems in which low-level interpretive processes are connected to a wide variety of sensors, and those processes in turn are integrated with higher-level systems that use recursive loops to perform more sophisticated cognitive activities such as drawing inference, developing proclivities, and making decisions that feed forward into actuators, which perform actions in the world. (24)

When operated by objects that sense (plants, animals, machines), the cognitive nonconscious forms an assemblage of processes deciphering, sorting, and cataloging available data, connecting it to the sensing technologies that are propelling the next wave of technogenesis via datafication. The implications of the cognitive nonconscious are crucial to include here due to their "ability to pose new kinds of challenges not just to rationality but to consciousness in general, including the experience of selfhood, the power of reason, and the evolutionary costs and systemic blindnesses of consciousness" (Hayles 2014, 199). Algorithmic technologies begin to predict the conditions with(in) which we perform our daily selves. By predicting, they gain the potential to sculpt the environment in/with/through which a human being's perceptual apparatus functions. Algorithms gain the capacity to shape the role a human being performs by suiting the environment to pre-planned situations. The formulation of the cognitive nonconscious creates a discourse for discussing the enmeshment between digital processes and consciousness.

When discussed in terms of technological assemblages of machine and human, we can begin to understand how thinking machines gain *"abilities to interact with humans as actors within cognitive assemblages"* (Hayles 2017, 174, italics in original). These machines use their advanced speed and cognitive agency to act upon the human "as presuppositions preceding sensation, as stimuli producing sensations and perceptions, as input through somatic markers into the cognitive nonconscious, and as experiences within the modes of awareness, consciousness and the unconscious" (174). Put more simply, computational sensing technologies can interact with individuals in temporal frames so quick that the perceptual apparatus doesn't even perceive an influence. This is the foundational basis for why data determinism has the capacity to reshape one's sense of self invisibly. In Hayles (1999) earlier work on the posthuman, she states, "We only see what our systemic organization allows us to see. The environment merely *triggers* changes determined by the systems on structural properties" (11, italics in original). When the environment of worldly sensibility is overrun with machines exuding increasingly overmatched levels of agency, the reality that is visible to human beings has the potential for becoming preselected and predetermined by those very machines through feedforward. Feedforward changes the feedback cycle of autopoiesis by filtering the input into a dynamic that fits the expectations/parameters of the machine. Within autopoietic systems, feedback loops create cycles of meaning-making in the *now* of being. Feedforward loops create meaning in the *to be* of being. Both humans and machines read information simultaneously as a form of nonconscious cognition; each then responds to this information through the process of sorting and filtering. Machine cognition, however, can respond faster and with more fidelity.

Feedforward is potentially difficult to manage because it implies that algorithms running smart machines are exponentially faster at processing and responding to the data generated within the loop, predicting what and how we should think. Because human consciousness takes approximately

500 milliseconds (half a second) to filter raw data into what we are aware of, it always delivers human reality later than machines can. The human unconscious operates at around the 200-millisecond range. For example, when I first began writing the early drafts of this chapter, I took a break to go to the grocery store. While walking up the street, I encountered two kids playing soccer on the sidewalk. I was deep in thought, thinking about how to explain neural networks, and therefore my conscious mind was in another form of reality. One of the kids made an errant kick that sent the ball toward the street. Without "thinking," I jumped toward the ball (it was only a couple feet away from me) and kicked it back to them. My unconscious mind saw the ball heading toward the street, and my cognitive nonconscious calculated the variables of what would happen if it made it out into traffic and told me to react appropriately. I had kicked the ball back before even consciously recognizing (being aware) that the ball was there. The time it took my cognitive nonconscious faculties to process that information was probably less than 25 milliseconds, the reaction began around the 200-millisecond range, and the act was only made real to my conscious self roughly 300 milliseconds later. The time between the cognitive process and the worlding process of consciousness was effectively lost to my aware self.

Citing neuroscientists Benjamin Libet and Antonio Damasio, Hayles (2017) refers to this difference between human cognitive perception and conscious realization as the missing half-second, which is "the temporal gap between brain activation and awareness" (190). Because non-biological sensing technologies like algorithms only operate in the speed of cognition, not consciousness, they can transmit more information inside the system of techno-performativity, immediately gaining an upper hand on their biological counterpart. Hayles cites the speed of the high-frequency-trade algorithms that run most stock trades as operating at the speed of "five milliseconds or less" (131), with the fastest executing trades at the 129-microsecond range (169). A microsecond is one-millionth of a second! The speed an algorithm can operate at, in conjunction with the speed of the intelligent hardware that runs them, completely upends the relational process between sensed information and human beings.

The algorithmic system running Google's search function for the definition of microsecond only took 339 milliseconds to deliver 30,500,000 instances of the term on the internet. Consider, part of that computational speed depended on the speed of the Wi-Fi I was connected to, as well as the computational power of my laptop, and the distance to the nearest Google data processing hub. In 2017, Google created the proprietary TPU (Tensor Processing Unit) 2 chipset used to first power its Deep Mind AI platform. The chipset bundled 64 chips in a processor array and could process information at 11.5 petaflops per second. That is 11×10^{15}. That means TPU2 can process 11.5 quadrillion data points per second.[1] In 2023, with the fourth generation of the TPU system, the system architecture has become even more powerful. The current system bundles together pods of 4,096 chips that have the theoretical capacity to process 1.1 exaflops or 1.1 quintillion data points per second (Jouppi and Patterson, 2023). With this processing power, it makes me question what amount of agency human beings ultimately have when engaged in processes of data determinism with intelligent machines. What power they may have lies in the fact that interaction with data sensing machines is a mode of both conscious and nonconscious techno-performativity that operates first as a form of user-initiated role play.

As will be shown later, role play allows us to face the challenge of machine learning head-on and requires us to "utilize the affordances of the very technologies that are responsible for marginalizing our sensory experience" (Hansen 2015, 4). Experiences developed under the Architecture of Role Play that highlight and/or harness a spectator's position as (co)author are a prime place to uncover and discuss the affordances. Therefore, it is important to better understand how role play operates in relation to feedforward cycles and how to analyze it as a mode of spectatorship using the framework of data determinism. Before covering role play and its relationship to data determinism, we should first cover a brief overview of algorithms and their deployment as the backbone for the various technological tools and systems that make up the infrastructure that allows this powerful technic to operate.

Surveilling Machines and Data-Based Systems

Today's paradigm of data determinism is based on what Ed Finn (2017) describes as a "culture of computation" (16) primed to collect, calculate, and predict the very essence of twenty-first-century human life.[2] Largely beginning to take root in the early 2010s, the internet and all its device entanglements began to focus on the "quantification of everything," leading to a "slew of new businesses that add an algorithmic layer over previously stable cultural spaces" (124). Led by companies such as Google/Alphabet, Facebook/Meta, and Amazon, data quickly became the currency of the twenty-first century, and the users who generate that data became the product. In what began as a supposedly benign process of assisting users to connect and learn about the world around them, these platforms began to morph into collection sites that could amass an incomprehensible amount of user information. With all this information on hand, Big Tech (and governmental agencies) gained the power to switch from being simple indexers of cultural knowledge to the very architects of culture (156). By interfacing with these algorithmically operated platforms for "free," users became the source of a constant stream of revenue based on targeted advertising. In the data economy, if a digital product seems free, it is because the user is what is for sale.

The above platforms allow data determinism to work by creating more personalized engagement for the user. More engagement means the platforms running algorithms gain one's attention for longer amounts of time with deeper investment; these are two of the primary activators of the experience economy which has morphed into the attention economy. For example, on social media sites like Facebook, TikTok, and Twitter (X), filtering algorithms feed us content that is calculated to grab our attention based on content from our previous interactions and the interactions of our friends' networks. The system often filters out material that one might consider less engaging. Engagement is necessary for companies like Meta, Amazon, and Alphabet because the tailored advertisement that constitutes their financial bottom line is more likely to stick or be acted upon by an engaged user, and in the attention economy, every microsecond of engagement means more profit.

To stay relevant in the face of these tech behemoths, nearly all media platforms had to engage in the same data collection processes to stay solvent. This has led to the sorting of information sources based on the highest level of personal engagement and interaction. The earlier paradigm of unlimited choice and democratic freedom through interactivity found in Web 2.0 platforms of the 2000s gave way to a constant ecosystem based on datafication initiated to limit that choice. Cheney-Lippold (2017) explains that datafication marks the "transformation of part, if not most, of our lives into computable data" (11). Every tracked move, every search query, every purchase, and every social media interaction becomes a node in the datafied network of daily reality. Ignas Kalpokas (2019) argues, "Algorithmic systems capable of sorting and ranking choice options available to individuals and then continuously learning from immediate reactions and continuous behavior of such individuals in order to re-alter the landscape that individuals find themselves immersed in, become particularly potent governance mechanisms" (59). While users might believe they are given agency and ultimate freedom, they become beholden to the logic and influence of algorithmic technologies. The social system under datafication is one of constant data collection feeds a cycle of power relations between human users who generate data and the programs that analyze and then feedback a simplified array of new potentials. This has been allowed to proliferate unfettered because the process is wrapped up in the mantle of convenience while offering increasingly individualized outputs. In this process of simplification and individualization, users often become increasingly limited in their horizon of experiences, effectively governing their views of the world around them.

The algorithms and data-based processes discussed throughout this chapter are part of a field of computer science often discussed, utilizing the terms Artificial Intelligence (AI) and Machine Learning. When referring to AI, I will specifically discuss the systems of online infrastructures developed prior to the launch of Generative AI built off Large Language Models (LLM) that serve as the foundations for products like Chat GPT, Bard/Gemini, Claude, and LLaMA. I highlight this as those products are part of the datafication paradigm and the data determinism processes, but how they operate under these paradigms is yet to be fully understood due to their recent launch. As such, I will bracket any discussion of them for the Afterword of this project. Machine learning is an adaptive process of computation that generates predictions (abstractions/inferences) based on correlations between data sets. To be adaptive, the learning process requires both input (data) and instructions for gathering and filtering (algorithms). Leading researcher in machine learning, Ethan Alpaydin (2016), defines algorithms as "a sequence of instructions that are carried out to transform the input to the output" (16). The simplest way to understand an algorithm is an instruction set or recipe applied to calculate data toward a specific result. With an algorithm, you can "define any kind of relationship as a functional thing that unfolds in time and space and articulates itself by its actual behavior" (Kleber and Trojanowska 2019, 102). Like the physical environment that surrounds humans, the space for data read by algorithms is the info-rich environment of the internet and all the internet connected devices that make up what is called the Internet of Things (IoT), discussed later in this section. As a network of networks, the internet works as an assemblage of data and the pathways from which to access and process that data. Think of the internet both as a large digital data library as well as a networked processing unit similar to a scaled-up version of the human brain. It is an architecture in the sense of both structure and process, like the architectures of exchange offered in this project. It operates as a container for inputs while serving as the platform for processing those inputs into new outputs and subsequent inputs that shape unique digital realities and datafied subjectivities.

Algorithms perform their instructions in order to define essentially any object. These instructions—which are often based on inherent human biases—are set to seek out specific outcomes based on the media ecosystem the algorithm is programmed to surveille. For example, within an ecosystem based on constant targeted engagement with devices and content, the instruction might be tuned in this manner: If a user interacts with X more than five times, then push Y to their media channel. In real-world terms, for every Facebook post read or YouTube video viewed that is tagged as political, the platform's algorithm learns to push more political content to the user. More cute cat photos/videos liked, shared, or viewed leads to more posts with animals until the posts are narrowed down primarily to cats. Everything is tagged with metadata based on rudimentary classifications—think stereotypes—that help algorithms sift and sort through the information.

Just like how human consciousness creates social realities through filtering out extraneous information, smart machines need algorithmic instructions for making sense of the data available. When an algorithm is sophisticated enough to not only sort and catalog data but also learn from that data through recursive feedback loops, it becomes the beating heart/ and brain of machine learning and AI. In some capacity, an algorithm is the perceptual apparatus of a smart machine. Most AI algorithms are designed upon the model of the human brain, into complex neural networks. These networks simulate the parallel processing found between neurons and synapses. For example, the Google search engine is a purpose-built AI that scours the multiple data inferences throughout areas of the internet. This is done through the interrelated processes of web scraping, data mining, and machine reading, where "programs automatically surf the web and extract information" (Alpaydin 2016, 47) and then cluster this information into smaller connected clusters. Not only is the algorithm searching for and delivering requested information to the user, but every time one enters a request into the search bar, they add new data to a repository, which helps make the AI "smarter." Smarter is used here as a simplifying placeholder to explain the complex process of becoming more able to predict outputs and potential future inputs based on correlations/associations between data points. The more data points, the increased number of significant correlations. Alpaydin explains, "Generally more data means more information, and hence more data tends to decrease uncertainty" (80). While we typically think that Google Search is primarily a tool to help us find information about the world, it has been from the start a data aggregating algorithm building a smarter AI. Digital assistants like Siri, Alexa, and Cortana operate in the same manner as invisibly surveilling machines collecting new data; they just change the user interface from search bar to voice assistant. In a conversation with Kevin Kelly (2016) in 2002, before Google went public, Google's CEO Larry Page informed Kelly that Google wasn't simply building a search engine, it was building the most effective platform for creating AI (37). This is well before the advent of the generative AI we are grappling with today. All those search queries are stored into a vast database used to make it a smarter machine by adding a multitude of new data points from which to correlate new knowledge.

The process of accumulating data sets and analyzing them into more distinct and useful associations is called data mining. Through regression and correlation of these associations, algorithms write the rules for everyday life. Like humans do, learning algorithms teach themselves by accumulating more and more precise bits of data and data about data called metadata. Metadata is a sub categorization of a larger data set. Metadata allows the system to breakdown any unique data point—a document or webpage, for instance—into smaller parameters. For instance, the document I am writing now has multiple parameters the program uses to identify what it contains and how it was produced. For instance, while I never explicitly added this information while writing, the metadata shows my name, the time the file was created, and last saved version, what iteration of the document I am on, how much time I've spent editing it, and a breakdown of the pages, paragraphs, words, sentences, characters, etc. The Google search bar has its own metadata related to your geographical location, time, machine address, and specifications, as well as previous searches. The search function operates by matching that personalized metadata to other metadata embedded in every web instance it sifts through. In the processes of machine learning and data correlation, metadata is sticky in that it has unique properties that connect it to other similar data points during analysis. As these algorithms learn to extract more raw elements from data environments, they create models which can then enforce predictive behaviors, both human and nonhuman.

As an algorithm "learns" to correlate more data, it also begins to limit possibilities, and this is what makes them smarter in the sense of surer of their calculations based on the limits. A world with infinite possibilities is ultimately incomprehensible to both machines and humans; this is what makes the need for increased data and limits on that data necessary. This seems like an infinite cycle, but to make the comprehensible world smaller, algorithmic machines will forever attempt to expand their limits while at the same time categorizing data into narrower data-types. The infinitude of this task is what will hopefully keep us from ever having truly sentient technology through what has been referred to as a singularity. In the singularity, AI gains human level consciousness and no longer requires new input from humans at all. Though often a fear within the realm of science fiction, this is one of the worries plaguing researchers, politicians, and everyday people in the advanced era of generative AI. We are not there. Yet.

The basis of the most powerful machine learning tools are deep neural networks modeled after the systematic processing of the human brain. Neural network projects such as Google's Deep Mind / Tensor Flow, IBM's Watson, and Amazon's Deep Scalable Sparse Tensor Network Engine (DSSTNE) set the foundation for cloud-based machine learning to push the limits of what was possible with technology. These networks use a complex system of parallel processes where vast arrays of artificial "neurons" are taught how to infer information without human assistance. In complex processes of datafication and machine learning, the "data starts to drive the operation; it is not the programmers anymore but the data itself that defines what to do next" (Alpaydin 2016, 11). Data and the smart machines that process that data begin to develop a symbiotic and closed loop relationship that no longer needs instructions per se, just more data. Learning comes by way of association. Humans understand the difference between a dog and a cat, or hot and cold, by comparing one against the other. Using regression, algorithms sift through millions of data instances to filter out anomalies and outliers, eventually creating norms based on patterns. For example, in a face recognition algorithm, the system is programmed (trained) to recognize certain attributes that make up a face: Shadow, lines, contours, colors, and distance markers between a nose and eye, for instance. To identify any specific individual, the algorithm simply needs enough existing data. Once it can identify one face, it learns to differentiate between two faces. My face is not your face. Every time someone uploads a new selfie or family photo to a web interface, such as Facebook, Instagram, or Apple Photos, the platform's algorithm gains a new data inference to compare against. After enough images, these algorithms learn to recognize the difference between faces. In a 2016 article for NPR, Naomi Lachance (2016) explains that Facebook's facial recognition software was considered at the time better than the one used by the FBI. Facebook's advantage was based on the multiple billions of worldwide users who consistently upload new photos to the platform which become data points to reference. Those photos are analyzed and tagged with metadata to help the company's algorithms learn. Lachance (2016) explains, "with its huge database of images, Facebook's algorithm has a leg up on most others in that it is constantly being taught how to improve. Every time you tag a photo, you're adding to an enormous, user-driven wealth of knowledge and data." As these platforms grow based on user numbers, they become exponentially more intelligent. Following this example, it is our actions that are helping make the systems smarter without them needing new programming to become so. The success of facial recognition is due to the sheer amount of data generated. The sophistication of current facial recognition arises because so many people are uploading images to their platforms. Photos are not the only data. The devices that run these algorithmic systems are collecting voice data, location data, engagement data, user pattern data, purchase trend data, emotional response data, political affiliation data, etc. This data leads to further advances in voice recognition, data transcription, affective computing, and today's generative AI. Though not discussed as often, affective computing tools also use algorithmic processing to track and learn human emotional responses based on behavioral characteristics such as "dynamics of signature, voice, gait, and keystroke" (Alpaydin 2016, 66). Add these factors to other biometrics, such as subtle differences in facial expression, and the machine can learn the intricacies of human affective states.

In the earliest iterations of machine learning, programmers needed to add a governing rule set to the system called a supervisor to keep the data inferences in check. Without this, the systems might make incorrect guesses based on noise in the system. Today's best learning algorithms no longer need supervisors. The sensors in an iDevice, for example, generate data from people's actions. The data freely or unknowingly given, such as the tagged photo upload or GPS location, automatically generates verifiable data instances from which to self-correct and therefore helps algorithms learn on their own. The capacity of deep neural networks to become autonomous has taken quite some time to evolve. The primary barriers were programming tools and hardware for both processing and storage. Because AI computation takes enormous amounts of individual processors, neural networks that could run in parallel required large "super computers" which were made up of smaller clusters of processing units, each programmed for specific tasks that could then be compared against each other for regression and abstraction. In the mid-2000s, researchers realized that the Graphics Processing Unit (GPU) chip, created to run images on video game systems, could simplify this process and allow smaller processing units to run in parallel. Today's deep neural networks are primarily built by tethering GPU and CPU units into networks running the same task multiple times against previously processed information into finer and finer correlations. As introduced earlier, Google's TPU was developed specifically for deep neural networks built into cloud computing infrastructures and is performing speeds exponentially faster than the previous GPU clusters used.

The next hurdle was storage capacity. Alpaydin states, "The earliest hard disks for personal computers had a capacity of five megabytes; now a typical computer comes with five hundred gigabytes—this is one-hundred thousand times more storage capacity in roughly thirty years" (142–143). The average desktop computer in 1998—the year I first owned one—had around only 6 GB of ROM (read only memory) and around 64 MB of RAM (random access memory). Today's iPhone is configurable with up to 1 TB of ROM and has a standard 8 GB of RAM, but most of our data isn't even stored or processed on our home computers anymore or even our smartphones. Most lives in the cloud: "A person no longer has one personal computer that holds all their data and does all their processing; instead, their data is stored in the 'cloud,' in some remote offsite data center, but in such a way as to be accessible from all their smart devices, each of which accesses the part it needs" (143). Cloud computing is the backbone of machine learning as the primary data repository and the network used to process much of that data. As more and more individuals begin to rely on Google Drive, Dropbox, iCloud, and Amazon Web Services for the storage of their data, the machines storing and processing that data become smarter and smarter. The next step in the process was to create more products connected to the Cloud for both data collection and data processing, allowing AI to emerge out of the multitude of correlations and abstractions.

In the quest to expand the limits of comprehension for machine cognition, more objects are being embedded with machine learning technology. All these technological objects are data hungry. This is one of the reasons for the push to expand the IoT. McKinsey Analytics (Chui, Collins and Patel 2021) defines the IoT as "sensors and actuators connected by networks to computing systems. These systems can monitor or manage the health and actions of connected objects and machines. Connected sensors can also monitor the natural world, people, and animals." This term often refers to the effort to make all objects "smart" and interconnected. Computer chips and the algorithms that run them are quickly being embedded in virtually every object imaginable. 2008 was the first year that there were as many devices connected to the internet as people at roughly 7 billion, and in 2016 that number had multiplied to over 24 billion (Hayles 2016). The IoT is relatively smaller than that number, with approximately 15 billion in 2023 and projected to double by 2030, but these devices are not just connected to the internet; they are designed to transmit data back and forth with each other. With the advances in miniaturization, software technologies, deep neural networks, and their impact on machine learning and AI, nearly any object will relatively soon have the capability to be connected and listening. Per McKinsey (Chu et al., 2021), the estimated global economic impact of IoT will range between $5.5 trillion and $12 trillion in the year 2030. These numbers were calculated prior to the advent of advanced generative AI. I list these figures to show the quantifiable impact of "smart" technologies as they become increasingly embedded and invisible.

The early IoT was first implemented to help humans work more efficiently when operating machines. Returning to Kelly (2016), he explains how some of the first commercial uses of data tracking chips were implemented in automobiles, marking an early version of IoT: "Every car manufactured since 2006 contains a tiny OBD (on board diagnostic) chip mounted under the dashboard. This chip records how one uses their car. It tracks the number of miles driven, at what speed, times of sudden braking, speed of turns, and gas mileage. This data was originally designed to help repair the car" (261), as the value of this data became more understood, it became more integral in other places. For example, in 2011, I connected my insurance company Progressive's "Snapshot" module to the OBD reader input on my Jeep which allowed the company to track my driving habits. After six months of tracking, I mailed back the device to the company and received a reduction in my premium based on the personalized data. This "snapshot" acted as a digital double, showing how I performed as my driving self. The data helped the company create a digital representation of my driving behavior. I am sure the company also shared this data with other corporate entities to gain a unique profile of my location-based behavior. There are multiple other uses for this type of data, such as taxing "drivers based on which roads they use and how often. These usage charges could be thought of as virtual tolls or automatic taxation" (251). Kelly identified 25 unique instances of large scale dataveillance we interact with daily:

> Car movements, highway traffic, ride-share services, long distance travel, drone surveillance, postal mail, utilities, cell phone location and call logs, civic cameras, commercial and private space surveillance, home surveillance, smart home integration, internet connected personal interactive devices, grocery loyalty cards, e-retailers, the IRS, credit and bank cards, e-wallets and e-banks, photo face recognition, web activity (cookies), social media, web browsers, book reading (kindle can track your reading progress and patterns), and fitness trackers. (255)

Consider today the variety of other ways the IoT gathers data from you. As more smart connections come online, the range of instances will most likely expand into all aspects of human life. Humanity is still in an early phase of the IoT, yet there is already enough data stored in the cloud to populate more than "320 libraries of Alexandria per person on the planet" (257). The way things are tracked today is much more embedded in daily life as well. For instance, instead of visibly having to plug into the ODB interface discussed above, that same technology is simply downloaded as an app on my smartphone, which supplies data simply through all trackable daily actions. It runs in the background and rarely proactively seeks my input, unless I have disconnected the data stream.

Much of the data tracked that is visible is marketed as beneficial to people. When the data is transparent and accessible, it offers us a form of role-play and authorship based upon constructions of a *quantified self*. This term refers to the ability to use a variety of sensors to track and record data variables related to a person's body and how this body performs in the world. This form of tracking is marketed as a way for people to monitor themselves with the purpose of using their data to selectively improve their daily performance. These sensors range from biometrics (heart rate, sleep cycles, temperature, oxygen levels) to location and motion tracking devices (Global Positioning Systems and accelerometers).[3] Wearable devices such as the Fitbit and the Apple Watch contain these sensors to give the user interactive user data marketed for self-health improvement. These sensors are also increasingly being embedded in our non-health specific everyday devices. As early as the iPhone 7, released in September 2016, each device contained a barometer, 3 axis gyroscope, an accelerometer, a proximity sensor, an ambient light sensor, a biometric fingerprint scanner, and location devices such as a digital compass, Wi-Fi, cellular receivers, GPS and GLONASS (Russian equivalent to GPS), and a proprietary technology called iBeacon. While largely unutilized today, this last technology could track its user's location down to mere feet if in an environment with Radio-Frequency-Identification (RFID) transmitting technology. iBeacon and the early beacon technology have been revived in AirTag technologies which serve as portable surveillance devices, used to track items such as luggage, vehicles, or, in nefarious instances people. Today's iPhone 15 Pro expands the GPS capability, adds an ultra-wideband chip for instantaneous near field communication, and has a LIDAR scanner that allows it to optically map physical objects, allowing the implementation of advanced augmented reality and digital replication of physical surfaces interacted with. All of these embedded sensors allow the iPhone (and similar iDevices) to track, record, receive, and transmit micro-data about its user constantly. As one of the primary nodes in the IoT, the iDevice is a constantly surveilling digital companion feeding data to centralized hubs that allow exponentially increased intelligence to grow in machines operating through deep learning and neural networks. Due to these invisible sensing technologies, lives entrenched in datafication "have become [both] data-driven and transformed into data" (Ilter 2017, 81). In some capacity, these devices become duplicate perceptual apparatuses.

The foundation of the IoT is to develop a deep web of dataveillance where all aspects of human life can be turned into data points for classification and calculation. The reach of dataveillance and datafication is profound and will lead to a wave of social change and upheaval. Internet researcher Sam Greengard (2015) connects dataveillance to the IoT when he warns:

The Internet of Things isn't just about locating objects and using them to sense the surrounding environment—or accomplish automated tasks. It's a way to monitor, measure, and understand the perpetual motion of the world and the things we do. The ability to peer into the spaces between objects, people, and other things, is just as profound as the objects themselves. The data generated by the IoT will provide deep insights into physical relationships, human behavior, and even the physics of our planet and universe. (169)

The risk in dataveillance comes from the process machine learning enacts: Correlation and simplification. This process enables smart machines the "ability of reorienting, or nudging, individuals' future behavior by means of four classes of actions: 'recorded observation'; 'identification and tracking'; 'analytical intervention'; and 'behavioral manipulation'" (Esposti 2014). While I do not address this four-stage cycle explicitly in this chapter, its relation to the loop of performativity and autopoiesis is apparent. Through the process of simplification, smart machines become more focused and narrower in the information they feed back and forward into the cycle. Humans are complex, but in their relationship with machines under datafication, complexities are intentionally eroded because they don't make "sense" to algorithms. With the trend of dataveillance expanding at an exponential rate, we must ask ourselves what is the best loop created in the interchange between smart machines and mediatized spectators? The recursive cycle used to make machines smart has the potential to make us less so by a constant process of filtering out data that doesn't fit in the statistical model that it believes its user to be. Recognizing this power is the first step in a mediatized conception of relations with data, the next step is asking what relationship we want with our machines. Kelly (2016) states, "Our central choice now is: What kind of tracking do we want? Do we want a one-way panopticon, where they know about us, but we know nothing about them? Or do we construct a mutual, transparent kind of 'coveillance' that involves watching the watchers" (259)? I argue for the latter. The Instances explored later both foreground the implicitness of the human in the act of dataveillance and also engage in strategies for considering a system of purposeful "coveillance."

Role Play: Playful Identification Inside and Outside the Ludic Frame

People and their data-determining technologies engage in an *agon* whereby each attempts to exert control over the other in terms of selfhood and identity creation. Play theorist Roger Caillois (1963) defines *agon* as an aspect of play that relies on skillful strategy between two or more opponents. It is a type of play with a potential winner. The players in the technogenetic paradigm of data determinism are algorithms and their human input generators. The game played focuses on the struggle over who has control over the shape of the person, and the primary mode of play is role-based. As explained in the first part of this chapter, this oppositional play is rigged in favor of the technologies, partially because of their speed and partially because of their hiddenness. Role Play is then an architecture through which mediatized spectators have the potential to reverse this dynamic when manipulating the data flow by playing with their own constructions of identity.

This last part must first take a bit of a departure from its corresponding structural components in the previous chapters. This is because role play is itself a social condition that can and has been examined divorced from concepts of both performance writ large and mediatization in specific. To define the *Architecture of Role Play* as a structure that can be read through the lens of data determinism, I must first briefly explain the relationship between the self prior to formal performances experiences and the self in process as a mode of performance. Following the theories of both play and sociology (Goffman 1961, 1959; Huizinga 1955), role play is a crucial aspect of mental cohesion and social understanding of the self as an organized construct in the social world. People are material beings defined by their perceptual apparatus, but the self is an assemblage of layers mapped onto that material substrate and only understood as a being in relation to the environments that the apparatus encounters. As social animals, people are structured in a way that is pre-conditioned toward acts of role-play where the multiplied layers of selfhood performed are reconfigured based on the socio-cultural situation one performs with(in).

Ervin Goffman is famous for introducing the idea of performing the self as a quotidian act of daily life that is theatricalized. Goffman (1959) explains that this act of self-performance must first be understood as a division of the individual first into character and performer. Goffman uses the metaphor of the stage as a place through which the performer (a self prior to characterization) takes on the role of character (an individualized, contextualized, and actualized self) to be witnessed by the audience within the performance of daily life situations. Here, the character that is witnessed is then associated as the true self to the witnesses who are also performing their own characters defined by the social situation. Thus, a true, original, and individuated self could only exist within a theoretical vacuum where there are no outside interactors to shape the performer's actualization of character. This theoretical space is inherently impossible, and therefore, we as individuals are *always* in performance mode playing our specific social roles. By introducing performance and character into the calculus, the concept of roles becomes a central aspect of one's own understanding of the self as a being always in process. Goffman (1961) uses social roles as the dominant mode of characterization within the theatrical model. These are not inherently fictitious roles (like on a real theatrical stage) but instead are flexible and stratified layers of selfhood defined by social determinates. Here, he argues that roles are the "basic unit of socialization" (87). Each role expresses specific attributes defined by the socially normalized understanding of the character. For instance, he references the judge who must be "deliberate and sober" and the bookkeeper who must be "accurate and neat" (87). These individual characteristics, traits, and attributes then operate as the ideal signifiers through which the individual gains the ability to access a framework for variability within the ever-shifting social frames they perform within. The self is therefore only a relational aspect of the individual always "involved in more than one system or pattern" that informs what its outward role must be (90). Due to this constant sense of re-shaping to meet the frame of the situation (the system/ pattern), the self should then be considered a process and not a stable object or material. The self is therefore a constantly changing set of performed roles put into relationality with its environment and then performed by the individual to match the situation.

In an era of deep mediatization, the material aspects of our data environment have become a crucial part of the social situation that causes those roles to always be in relation. As Couldry and Hepp (2017) explain, "Data processes here enter the fabric of the self's reflexivity" (154). Digital data is just one aspect of the wider mediatized social frame, but it becomes an overt and powerful determinate of what roles can be played due to the way it algorithmically shapes one's environment. In the edited volume *Playful Identities* (Frissen et al. 2015), the authors focus on the relationship between our media environments and the increased visibility of our engagement with role play to argue that "The construction of identity has become a highly reflexive project, and communication media are at the very heart of this reflexivity" (35). Reflexivity is deeply bound up in the relational elements of world building as discussed by the various scholars of media and culture mentioned throughout this project. What differentiates the Architecture of Role Play from the everyday variety is how it foregrounds the necessity of consciously activating the capacity of character as a way of fitting within a specific ludic structure.

To those readers coming from the world of theatre and performance studies, role play as an activity probably needs little explanation. It is the enactment of character that Goffman describes but placed within a fictitious frame where the character is often dictated by the given circumstances of the narrative work written by the playwright or developed through the generative processing of devising. The character arises through an analysis of psychological and physiological attributes that the actor then applies as a layer over their own individual persona and does so not instinctually but purposefully in order to bring the drama into reality in front of the audience. In experiential spectatorship, the audience member becomes both the actor/performer and (co)author through their purposeful adoption of character creation through role play. As a mediatized spectator, a role-player gains the knowledge and power to (co)author their own self/identity in a purposeful manner that allows them to work with the system they are playing within. For example, in the previous section where *Pokémon Go* is referenced, I as the player (game play) could simply perform as the non-reflexive version of myself in play, acknowledging that there are aspects of social role play always occurring. I was not required in any way to become my avatar, Professor Cobrah, to catch the digital monsters, but I was still unconsciously playing the role of tourist in the Santa Fe Square. Similarly, in *Adventure One*, to have played as a stockbroker or an industry titan would have potentially changed one's relationship to the game, unless, of course, the player was a stockbroker or industry titan. In role play I would however purposefully adopt those character identities in order to play the game.

Similar to how the technic of virtuality attempts to simulate affective experiences through projections of space and time interfaces through bodily senses, data determinism offers discrete simulations of personhood. While mostly divorced from deep engagement with technoculture, Scott Magelssen (2014) utilizes the terminology of simming to define both the active action of role play and the live event that action takes place in his study of popular historical re-enactment sites. He defines simming as a "deliberate embodied practice" engaged "within a simulated, three-dimensional physical environment" (5). The site also becomes a simming, or place for simulating the performance of meaning-making, outside the frame of real-life. Unlike my own framing, Magelssen differentiates simming from other forms of role play that might occur in first-person interactive gaming and instead focuses on live events where the audience member takes on predetermined roles that suit the performance environment similarly to how an actor takes on the character of the play. In games studies, the term simulation is defined as "a procedural representation of aspects of 'reality'" (Salen and Zimmerman 2004). Magelssen's sims are much more aligned with the theatrical representation of historical realities as interactive performance. While the sim in his examples has the agency to disrupt the specificity of the character played and often presents the possibility of playing roles vastly different from the players outwardly visible identity, they often are clearly defined to meet the specific needs of the narrative that frames the event. This is analogous to playing a character in a pre-written dramatic play, where the agency to contribute largely is defined by how one embellishes the character through acting technique and not authorial control. While these examples often lead to meaningful experiences for the sims, they can be more restrictive than role play within ludic architectures. In some way, this brackets off the potential that comes out of the act of purposefully authoring the role oneself. While his references are useful for those specifically in theatre and performance studies, I prefer to reference role play experiences that are more individualized and offer the mediatized spectator the ability to access ludic creative modes of agency and exchange. For that reason, I'll frame my brief explanation of creative role play through a more adventurous approach, those coming from games and media studies may have more familiarity with. I'll describe role play through the primary mode of activity and exchange found in tabletop role playing games or TTRPGs for short.

In the introduction to his TTRPG *Bestial Acts* re-published in *Second Person: Role Playing and Story in Games and Playable Media*, games scholar Greg Costikyan (2007a) defines role-playing games in this manner.

> In a role-playing game, each player takes the part of a single character in an imaginary world. One person, the Gamemaster, acts as a combination of narrator, playwright, and referee. He creates a world and a story for his player to explore, providing background, emotional context, and main encounters. The players have complete freedom to determine how their characters respond. (349)

Costikyan's game is an odd outlier within the world of TTRPGs as it does not have a direct connection to fantasy, sci-fi, or worlds of heroes and antiheroes found in the lore of comics. Instead, his game is modeled after the theories of Bertolt Brecht and, as such, has both an intellectual bent and is geared toward socially conscious artmaking. He playfully states that the game "will probably sell in small numbers" because the game's intended purpose is "to take the paradigm of role-playing game and apply it to artistic effect" (349). What *Bestial Acts* has in common with the more popular types is how it is a game of imaginary action where the purposeful decisions of the role players determine the trajectory of the narrative in many ways. Without the co-authorship of the players in concert with the game mechanics, the experience could not exist as anything other than a loosely structured narrative. In these forms of role-playing games, there is a certain required level of give and take that allows the player agency to create and then insert their own fictitious persona into the world of the play in a manner that changes the world played.

In popular games such as *Dungeons & Dragons,* the intended purpose is based on entertainment, imaginary exploration, and community building. For example, I have been playing a *D&D* campaign with my partner and friends over Zoom for the past three years. In this campaign, *Curse of Strahd*, a narrative based on the *Ravenloft* series from the 1980s, we have formed together a group transported into a land covered in mist and populated with creatures from our nightmares. In this sprawling narrative, our team of adventurers must engage in a quest through the cursed land of Barovia to solve the question of what has happened to all its inhabitants under the rule of the vampire Strahd. Ultimately, it is part gothic horror and part choose-your-own-adventure story that leads to a battle against one of the most powerful undead villains in the history of *D&D*. Like all campaigns in this world, one of the key components is developing your own character that you will play throughout the adventure. My own character is a neutral-evil Tiefling warlock named Brand, who has an insatiable desire to eat the bones—and sometimes the faces—of all our vanquished enemies. He typically wears parts of the vanquished enemies as some form of totem. He also has a truly short attention span, an inability to work well in a group setting, and a propensity to light things on fire for no reason. As part of my own role play, Brand continually gets our team into trouble and often disrupts any attempts to strategize effectively due to playing the character's impulsive demeanor. I mention this due to the way it often breaks the system of the game developed by our gamemaster in concert with the narrative structure of the game. On more than one occasion, I have made a character decision that sends the planned encounter off the rails. As a character with ridiculously high deception traits, Brand often runs into a room/dungeon without giving the team advanced notice. Once inside, he will often encounter a horde of evil creatures ready to pounce. Because of the deception skills, I/he usually announces that he is some outlandish form of inspector. For instance, when dashing into a cave full of werewolves, I loudly declared that I was a game warden, given the mandate of cracking down on unlicensed hunts. This deceptive pronouncement was then checked with a d20 dice roll, which for my character always comes with a +13. In this example, a 32 was the final dice score, leading the werewolf leader scrambling to defend their pack instincts and then allowing me to set off a powerful fireball spell at an advantage, killing more than half of those in the room. This encounter should have lasted our group over an hour of combat but took less than five minutes. Our gamemaster is very flexible in letting us play the game in character and using our role in the most advantageous way. This follows Costikyan's (2007b) explanation for how *D&D* differentiated itself early on from other forms of tabletop strategy games: "If you could imagine it, and the gamemaster was willing to go along, it could happen. This opened an exciting vista of vastly more free-form and flexible games" (5). Without engaging in this form of ludic creativity and being allowed to do so, my actions and my character persona might simply be defined by the game's mechanics, an algorithm of sorts. By creatively authoring how I manipulate the character in the narrative, I have been able to "break" the game in many instances, more often than not, to the amusement of the rest of the team on the other end of the Zoom link. Here, playing the game not strictly based on the rules leads to a more enjoyable and meaningful experience for everyone. Role play is therefore a structure that allows the ultimate level of creative agency and flexibility to shape the experience in ways that allow new forms of meaning making.

The previous chapter introduced Eric Zimmerman's (2015) *Manifesto for a Ludic Century* as a way of explaining how the twenty-first century has become one in which all aspects of life become aligned with play, specifically games and their structuring of time and space. As I've explained throughout this chapter, the ludic relationship with our techno-environment is increasingly one of engagement with data. Zimmerman confirms this when he states:

> The ways that we work and communicate, research and learn, socialize and romance, conduct our finances and communicate with our governments, are all intimately intertwined with complex systems of information—in a way that could not have existed a few decades ago. […] When information is put at play, game-like experiences replace linear media. Media and culture in the Ludic Century is increasingly systemic, modular, customizable, and participatory. (21)

I previously cited Zimmerman to explore the ways that iDevices prompt an increased prevalence of gamification on mediatized subjectivity. The formal constraints of games are more objective based and bounded by specific rules defined by space, time, and place. In structures of datafication, many of these rulesets are thrown out the window, which allows constant learning from ever shifting inputs. Mediatized spectators, as co-authors, operate in these structures by manipulating the inputs, which then allows them to perform multiple roles and identities as a new form of output and possible data point that becomes difficult for the algorithmic systems to process efficiently.

As will be shown in the Instances to come, a purposeful creative authorship and enactment of character roles are necessary to seamlessly fit within the game and the narrative systems that allow them to operate best. One can, however, resist that "seamless" enactment by playing a different role that does not match the frame. This possibility of performing outside of role is an area I want to highlight as a possibility for confronting the unseen forces of data and operates as a form of techno-performativity. Following the logic of role-based hidden relationality that exists through the paradigm of and subsequently the technic of data determinism, I propose that we think of the mediatized spectator as one who can perform a playful identity that "has the quality to restructure itself according to the experiences one encounters" and "by engaging in role-playing, for example, one can see that multiple characters (or identities) can be explored and played out" (Deen, Schouten and Bekker 2015, 115). Through role play, mediatized spectators can perform a model of resistance to the technic of data determinism, gaining agency to develop multiple real selves. These selves can work against the data doubles operating as their digital identities developed within systems of datafication. For this reason, spectators may still have the power to manipulate these systems through active role play.

5.1   The Spectator as Flexible Character: (Co)Authoring Identity Through Role Play and Data Interactions

Algorithms and smart tech have the powerful potential to delimit human identity and subjectivity. The Instances offered here present a challenge for thinking anew relationships with(in) our digital and non-digital landscapes. In them, a new mode of agency arises where spectators become (co)authors through acts of role play as a system of creative ludic exchange. The Instances highlight the potential for manipulating systems of datafication when spectators understand their implicit place (role) with/in systems of data. In each, I discuss mediated experiences that can be read as investigations of and interventions against the process of data determinism. They offer a way of understanding how a mediatized spectator operating as a creative author might utilize the performative power of role-play to counteract the subjectivizing power of data-based systems. As examples of exchange with experiential media, each offers a mode of role rehearsal that can then be capitalized on when plugged back into the algorithmic system. Because of the fictitious framing of their media containers, they "push the structuring tensions between continuity and difference more to the extremes, making them easier to discern" (Matzner 2019, 127). The Instances do not necessarily show how data turns one into a mediatized spectator but rather how one becomes an author of their own selfhood through acts of role play. The materialist phenomenological perspective of datafication reminds us how we become subjects under datafied systems "unless, that is, we refuse to be" (Couldry and Hepp 2017, 142). The process of becoming (co)author highlights the possibility for resistance to happen once the spectator understands their implicit place (role) with/in systems of data. The response from a mediatized spectator comes in a variety of ways: 1) Passive acceptance or control, 2) Passive resistance (co-operation), or 3) Active Resistance (authorship). By staging the affects and effects of data determinism through the logic of role-play, each Instance brings to the forefront of a mediatized spectators' conscious mind the way they are entangled and embedded in playful paradigms of identity and self-construction. The datafied condition of everyday life instantiates a playful battle over control of one's own sense of self. The Instances discussed both stage this condition of life and show possibilities for playing one's way out of the confines of the database.

Instance: I Am Not Trevor – Opposing Dark Fantasies in *GTA 5* and *Red Dead Redemption 2*

Time to take a picture of myself for posterity. So here I am, strutting about nearly naked except for my stained tighty whiteys and aviator sunglasses, ready to create some trouble (Figure 5.1). That is, after all, what my avatar is predicated on, creating chaos out of pure id. My name is Trevor Phillips, and I am one of the most divisive protagonists in video game history. As a youth, my father abandoned me at a shopping mall. So, of course, I burned it down. I tried to join the Air Force, but they said I was mentally unstable, but not before teaching me to fly, so I turned that skill into a trade as a drug runner. In various missions I engage throughout Los Santos and the surrounding area. I will push the extremes of what is possible and plausible for a playable online character. How do you feel about torturing hostages? Do you get queasy about cannibalism? How about uncontrollable bouts of rage? Are you up for the challenge of embracing your dark side?

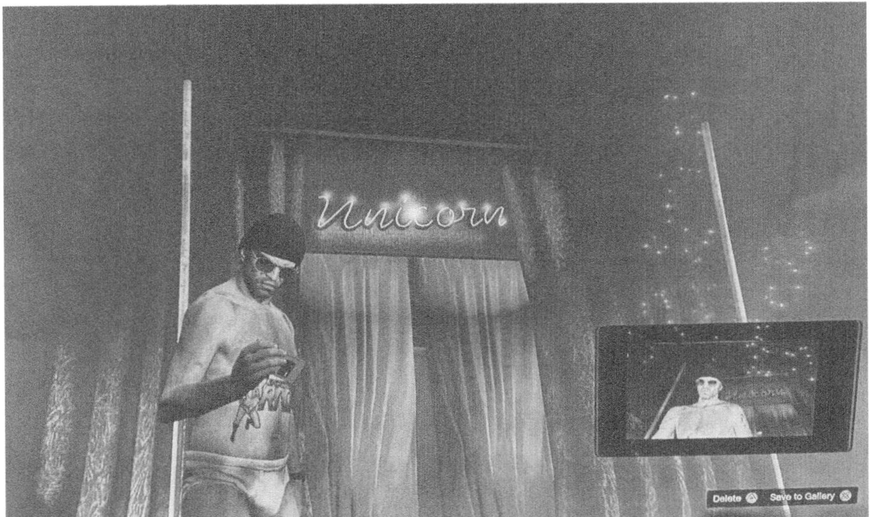

Figure 5.1 Trevor taking a selfie onstage at the Vanilla Unicorn in *GTA 5* by Rockstar Studios.

Trevor Phillips is one of the three lead protagonists in the wildly successful *Grand Theft Auto 5* developed by Rockstar North. First released in 2013, the game has sold the second most number of individual copies at 190 million, for an estimated 11 billion dollars in revenue (Takahashi 2023). This does not count for the multiple in game purchase options released for *GTA Online* launched two years later. Only *Minecraft* has been more lucrative a game in terms of the number of units sold. Trevor is scripted as an unruly psychopath created with the intention of giving players a protagonist who embraces all the antisocial and horrific capacities of the open world rampages built into the game. For this reason, he is often considered the epitome of an Edgelord which the Merriam-Webster dictionary describes as "someone who makes wildly dark and exaggerated statements (as on an internet forum) with the intent of shocking others." His actions and his narrative are intended to make you cringe. He is also the game's comic relief and the character often considered the representative of the stereotypical "average *GTA* player." This average player is often considered one who simply wants to rush around with an assault rifle, bazooka, or possibly a stolen tank and cause mayhem, break as many social rules and taboos, while garnering themselves a five-star wanted level. Trevor's narrative, backstory, and game path encourage all the above. This ability to role play the worst possible character traits all in one avatar has been one of the reasons the game has been both wildly successful and often criticized.

Below are a few excerpts from representative player posts in user forums related to Trevor. It is a testament to his polarizing character that a decade after the game was launched, and eight years after he disappeared from the online extension—killing him is a climax option of story mode—that endless forums on him continually pop up online. The overarching sentiment is that his character is constructed with a persona and character arc that matches the way players engage in the chaotic, illegal, and antisocial behaviors the game is best known for. He becomes the avatar that justifies the games engagement with the players' dark fantasies the game is often panned for by non-gamers.

GTA Forums – "Trevor Phillips: The Cringiest Edgelord in GTA History"

https://gtaforums.com/topic/981982-trevor-phillips-the-cringiest-edgelord-in-gta-history/#comment-1071841305

[FewPoleCat42] April 19, 2022
Trevor is a character with a very divided reputation. You either love Trevor, or you hate Trevor. There is no inbetween that I've ever seen.

Reddit – GTA SubReddit – "What is your opinion of Trevor Phillips?"

https://www.reddit.com/r/GTA/comments/119ccvx/what_is_your_opinion_
on_trevor_phillips/

[anon-deleted account]: Feb 2023
He's absolutely insane. Seriously, though, he was royally screwed up, but he was the most honest of the whole bunch. He didn't pretend to be a good guy or to have some better motives, he was what he was and he accepted that, and he was sincere and upfront about stuff and told people things like they were, without sugarcoating it.

[King Jehova]: Feb 2023
Great protagonist 👍 finally a GTA character who doesn't have any pesky morality that makes rampaging with them out of character.

GTA Forums – **"Trevor is an in-game representation of the average GTA player."**

https://gtaforums.com/topic/671256-trevor-is-an-in-game-representation-of-
the-average-gta-player/

[JustRob]: Jan 4, 2014
I think he is in fact the in-game image of what most GTA players are like.

[PedoneRosso]: Jan 4, 2014
Repost: Imho, Trevor is the correct personification of The GTA player: how else could one justify the kind of things you do when you play the games of the series?

[cp1dell]: Jan 4, 2014
Repost: Trevor was Rockstar's response to people who think GTA is a role-playing game, and complained that Niko had character during the story, but went against that during freeroam when **the player** did whatever they wanted to.

In many video games, one plays a hero of sorts whose objectives justify their actions through the game. For instance, if you are in search of a kidnapped princess in a fantasy game like *Zelda*, it justifies your vanquishing a variety of monsters along the way. Similarly, if playing a first-person shooter based on a real war dynamic like *Call of Duty*, the people killed during missions are framed as the enemy and thus must be taken out to complete the game. There is no real necessity for role play in these instances; objectives determine character. In previous versions of the *GTA* series, it was clear that there was tension between the dynamics of the characters and the actions one can take as the player. This became more apparent when the developers added first-person mode in a 2015 game update. Mark Walton and Peter Brown's critic's review explains:

> I'm here, mowing down wave after wave of police on the city streets, and for the first time while playing a *Grand Theft Auto* game, I feel immensely guilty about it. This isn't because of some grand moral awakening on my part, but an interesting side effect of what is the PlayStation 4 and Xbox One version of GTA V's most compelling new feature: first-person mode. Even when GTA games were top-down shooters, there was always something of a disconnect between the sometimes shockingly violent scenes on-screen, and the mentality of the player. You could imagine that, despite directly controlling a character, it was this virtual caricature of a criminal committing the crimes—you merely played witness to them. First-person mode fundamentally changes how you view GTA V's world. It has the power to make you stop and think about your actions, and to deeply question a character's motivations. And in a series that has long been criticized for glorifying a life of crime, rather than questioning it, this is no bad thing.
>
> (Gamespot.com)

This new mode offers a switch out of the avatar perspective where you can see Trevor running about the streets and instead puts you in his shoes, so to speak. By doing this, the question one might ask is, how does one justify the ability to simply jump in a car and run down pedestrians or rob people on the street if the character would not reasonably do so? As a game that mixes free roaming and mission-based play framed by crime narratives, the player often non-consciously takes on the role of maniac on the streets. In top-down and third-person mode, the player has an avatar as a mediator standing in between their controls and the actions on the screen. This is not true in first-person mode. In this mode, it is the player representing the actions and therefore, having Trevor's character well defined as a no-limits persona allows them to engage without having to defend their actions as anything other than outside the confines of character. A character would need to have some form of flaw that confirms the motives behind these actions for it to make sense in game, or it is simply a character trait of the player. This is what Trevor represents, a protagonist with no real moral compass—he does have a couple redeeming traits—that allows the player to reconcile their actions in the game beyond just simply "letting off steam" within a fictional universe. Trevor is one of the first true constructions in the *GTA* universe that can justify the players' in-game actions. But what happens when the player refuses to play along in Trevor's drama following his character traits?

Early in Trevor's character arc, he meets up with one of his hillbilly trailer park neighbors to do some target practice. This mini mission operates as a tutorial on sniper gun use and opens an optional side quest related to big game hunting for money and rewards.

Trevor Philips: Hunting? Hunting what? Retirees?
Cletus Ewing: Nah, just stuff. Windows, antennae, tires, vermin!

(Rockstar North)

As I play as Trevor, I/he follows Cletus around the trailer park until I/he finds the appropriate location to have some fun shooting up the neighborhood. Cletus instructs me/him to practice on the local satellite dishes.

Cletus Ewing: I'll introduce you to one of my all-time favorite pastimes. You're gonna love this. Okay, you see them three big satellite dishes? A decent shot and they'll go down easier than a whore's drawers.

(Rockstar North)

This seems simple enough, and it does seem like something Trevor would do for fun on a drunken Saturday night when he's running low on meth to smoke or fights to start for no reason. Actually, it seems like something he'd do just out of sheer boredom. I take a few minutes to learn the controls and zero in on three different dishes and expertly destroy them with a shot from the rifle. Eventually I master the controls and now on to more difficult targets.

Trevor Philips: So, what now?
Cletus Ewing: We're gonna try our luck on something less stationary.
Trevor Philips: Ooh, liberals?
Cletus Ewing: Nothin' as slippery as that. You'll see.

(Rockstar North)

Cletus leads me to a new perch and dares me to take aim as cars moving down the highway as a way of testing my ability to hit a moving target.

Cletus Ewing: I bet you never shot out the tires on a car before?
Trevor Philips: You'd be surprised.

(Rockstar North)

Again, this is entirely not out of character for Trevor. Causing havoc, and possibly even killing someone for no real reason makes sense in his alternate reality. For instance, when first introduced to Trevor in the game, he stomps a local biker to death for questioning his advances on the biker's girlfriend. This is a very sanitized retelling of this event; I'll spare you the gory details. So, simply shooting up a passing car is rather mundane and tame a task for Trevor. I take aim and shoot at a few cars, but am purposefully hitting the tires and nothing else, attempting to not inflict deadly harm on the NPC's traveling down the highway in some way veering away from Trevor's impulses. Upon completing this section of the mission, we travel by stolen golf cart to a nearby rundown motel for the final task.

Cletus Ewing:	Just two fellers killin' time in small-town America. Don't get much better than this, does it?
Trevor Philips:	The two most popular guys in town!
Cletus Ewing:	Am I glad I ran into you. Ain't nowhere near this fun bein' antisocial on your own.
Trevor Philips:	Whatever cranks your tractor, I say. What's up next on the hillbilly anarchist agenda?
Cletus Ewing:	Ever shot anything with a face? Actually… let me ask that again… ever shot anything with a face on four legs?
Trevor Philips:	Now, Cletus, it's important you realize I'm a man plagued by vicious rumors.

(Rockstar North)

Now Cletus wants Trevor/me to take out a pack of roaming coyotes who have been scavenging on the various heaps of trash scattered about the trailer park. When playing the game, my partner was watching and objected to this killing of animals for no real reason. After all the animals were just eating trash. This made me stop and think. For some reason, it does just strike me as unnecessary. Why go to the extremes of animal cruelty? I understand why this task exists in terms of the game mechanics, but the rest of the game really does not require other examples of murdering animals; the big game side missions are not mandatory. I know they are just digital renditions of furry creatures, but it separates my own persona from Trevor even in the first-person perspective, highlighting the ability to break free from the game should I want to. I try to justify not fulfilling the mission by imagining that Trevor would more likely capture and train a pack of wild coyotes than kill them. Wouldn't you be terrified by a person in control of a pack of "blood thirsty" beasts? At least that is what they would be after Trevor got a hold of them. Within the game, not shooting the coyotes means not completing the mission, but this is also a non-critical tutorial side quest that is not necessary to complete the primary narrative. By allowing me to sidestep the requirement to shoot the animals, it opens a minor moment of reflection on my role as a player within the confines of the digital coding. I was not the character defined by the narrative no matter how much Trevor's backstory and cut scenes tried to define me in that way. The tension created when recognizing this fact leads me to consider the other ways I might push back against real-life programming by (co)authoring my narrative.

In the online extension of *GTA 5,* multiple examples exist of players creatively remixing and reframing the landscape of Los Santos through its open world format. This is possible through a multitude of private role play servers that exist specifically for those who want to author their own narratives and experiences.[4] In these walled off digital universes, *GTA 5's* primary digital code serves as an experiential playground for the games "Director Mode," where you create your own persona and then act out their role. In the various servers, you might continue the mayhem that is typical of the game or simply role play the mundane by inhabiting the likes of a garbage man, corrections officer, or baker. Here, open play is an option, but the servers are often framed through host instructions to play as that character would normally when in the world. If they deviate, they can be kicked off the server. So, if playing a cop, going on a killing spree is out of character. While they do serve as a playground for identity construction, the servers are still rigid like many games due to these rules. A noteworthy example that pushes the confines of the game-world includes a word-for-word re-enactment of Shakespeare's classic *Hamlet* produced by the company Rustic Mascara, with a full cast of avatars playing out the story across different locations, including the top of a skyscraper and on a yacht. With the project having been recently the subject of a documentary titled *Grand Theft Hamlet,* the Covid inspired digital theatre creation is no longer available for free viewing on YouTube, but other examples of online role play in the crime landscape exist there. While giving the player freedom to explore in this *GTA* version of *Hamlet,* it still confines the players coming together in multi-player mode to follow a specific script with pre-written character traits. What these servers offer is a glimpse into the potentials of data-based role play (Lewis 2021) where the construction of digital alternate selves highlights the fluid nature of identity in the mediatized actual world where the rules can be blurrier.

Rockstar's outlaw game *Red Dead Redemption 2* integrates a similar player mechanic found in *GTA 5.* The difference is how the game allows role play to shape the system based on how they want to interpret the protagonist's character traits. Following the previous Trevor example, quite a bit of killing of animals occurs in *RDR2,* but this is largely justified as the protagonist lives in the Wild West of the late 1800s. They make money from selling furs and meat while also maintaining their health by eating what they catch and kill. They also can level up by making clothing and camp materials out of the skins. This game mechanic justifies some of the actions related to animals, but it is up to the player to justify their actions related to crime. Like the *GTA* series, the lead character Arthur Morgan is a member of a notorious gang of bandits who rob trains, steal wagons and horses, and complete bank heists. He can also indiscriminately kill NPCs during these actions if that is the player's prerogative. The game, however, introduces what it labels as an Honor System as a way of complicating the player experience.

This system introduces the question, "Do players want a morally upstanding outlaw who does bad for good reasons, or a heartless killer who revels in the crimes he commits?" (Prater 2023). Arthur commits all types of bad deeds, but unlike Trevor in *GTA 5,* the player has the option to commit good deeds, which makes their progress through the story easier in terms of rewards given. While the rewards are plentiful when playing good, it is also the more difficult option. If you choose to aim for high honor, it also helps to frame Arthur's story as one of old west heroism in the face of new world progress versus simply one of dastardly vigilantism.

The ability to role play Arthur in this way does not significantly change the outcomes of the narrative but does impact the player's perspective on his final outcome and possibly the narrative frame of his death. At the end he can die peacefully or be viciously murdered by a turncoat member of his gang. The honor system helps steer toward one version or the other. The system however has no impact on the fact that he will die due to an encounter earlier in the story. Mid-way through the game, the player/Arthur contracts tuberculosis while on a mission to retrieve money loaned by a local farmer. This scene offers three levels of aggression that lead to minimal or extreme violence against the farmer. The only way to not get TB is to stop playing the mission. The game does not end by doing so—there are plenty of other things to do in the open world—but by not collecting the money, the main narrative will not progress. This offers an interesting experiential possibility to the player, stay good (or bad) and live forever in a world largely without real purpose beyond paidic survival, or submit to the disease while pushing forward in the story toward Arthur's incredibly moving demise. Returning to the honor system, should you choose to break bad, his death seems to feel more warranted, and it is easy to move on. When I played Arthur in the first playthrough, I tended to dig into his good side as best as possible, keeping the Honor system on the positive side as it seemed more in line with Arthur's motivations and intentions based on backstory and narrative. This led me to be deeply saddened and feel empathy for Arthur when he succumbed to the disease in the final act. When playing in this honorable manner his death is one of hopeful longing as he watches the sunrise while succumbing to the disease. By playing honorably, I had a deeper sense of affinity with his plight. It also made me feel much more upset anytime one of my/his horses died, either in a mission or by some errant mistake in game play. Rest in peace, Thor, my beloved grey shire workhorse.

In both instances of role playing the characters in these games by Rockstar Studios, I am made to consider the ways in which one either leans into or against the algorithmically programmed character traits of the protagonists. By doing so, I am also introduced to some of the difficult to manage aspects of data determinism and (co)authorship in ludic frames. I gain the ability to work both with and against a system developed to control my actions toward specific outcomes. The outcome will be largely fixed within these games, but in the supposedly open world game of life, there is an endless number of potentials always ready to experience based on the role I am willing to author in the moment of interaction with both the people I face and the machines I use. Purposefully knowing there is a possibility to play and create however one chooses is what one learns through online and ludic experiences of role play.

Instance: Resisting Data Determinism – Multiple Roles/Multiple Perspectives

So here I am, hiding behind a dumpster in a dirty alley in downtown Brighton. I'm doing my best to maintain my cover and make sure that the two people fifty yards ahead are not aware that they are being watched. I have been following these two for the past ten minutes since they slinked out of the pub. Obviously, they must have been trying to give my team the slip, knowing that we were getting too close to uncovering the truth. What can they be talking about, I wonder, and what is it that they are scheming? Something horrible, I suspect. They are political extremists, after all. With each move through the neighborhood, I begin to distrust them even more because they seemingly are not doing anything that should raise any suspicions. I've been tasked with finding out what cruel intentions Alice has up her sleeves, but all I have been able to tell so far is that she and her companion are having a leisurely stroll on a warm summer evening, telling jokes, and catching a moment to smoke a cigarette together. Surely there is more to this encounter; I've been told they are up to no good. Is my distrust unfounded?

In the above experience, I was performing the role of covert agent in the production *Operation Black Antler*. I had been tasked with profiling the behavior patterns of supposed political extremists in an experiential performance based on role play. As part of this role play, levels of distrust, fear, and contempt for the political "Other" were amplified to justify my own actions. The experience offered me a sense of the ways algorithms embedded in digital media interactions condition my beliefs and daily behavior. It also caused me to question the ways these algorithms invisibly drive political messaging that determines how we perceive and behave toward those around us. Specifically, those of different political beliefs. It is in data determinism that our politics become primary identifiers, framing entire social realities and the ways we perform within these frames. In an algorithmic era where every digital interaction is tracked, analyzed, and recorded to manipulate our future behavior, it makes me question how we might resist?

Operation Black Antler (2016–2019), co-created by Blast Theory and Hydrocracker, is described as an immersive performance experience where the spectator is invited to "enter the murky world of undercover surveillance and question the morality of state-sanctioned spying" (Blast Theory n.d.a). Blast Theory describes their work as one that "places the public at the centre of unusual and sometimes unsettling experiences, to create new perspectives and open up the possibility of change" (Blast Theory n.d.b). Hydrocracker develops work that uses site-specific and immersive aesthetics to tell "stories that challenge what is thought and believed about contemporary issues" (Hydrocracker n.d.). Both companies utilize audience experience as a way of negotiating some of the most difficult to discuss political topics. The production began as a reflection upon the discovery of unethical standards used in the process of undercover policing in the United Kingdom and operates as a performative exploration of the ethics of state surveillance while utilizing the same tactics used by those who surveille. In this manner, the role-playing spectators are programmed to perform as algorithms themselves.

The spectators utilize role-play to encounter political opinions and viewpoints that are often filtered away from their daily routines due to algorithmic processes. The project effectively questions and challenges individual biases by encouraging an encounter with those outside their political bubbles using performance. In personal interviews with Matt Adams of Blast Theory, he explains that "there is a possibility in any exchange where two people come together." This is the emphasis behind much of Blast Theory's work over the past 30 years and particularly true of *Operation Black Antler*. Adams explains further that they "are interested in how you create conversations with strangers. How do we create an experience where even though we do not know each other? We can create something rich and powerful. Meaningful to both you and me, the maker of the work or architect of exchange."

The exchange offered in *Operation Black Antler* uses deductive game mechanics in an algorithmic fashion, asking individuals to surveille, correlate, and determine future police action based on conversations with characters who exhibit supposed ultra-conservative ideological viewpoints. This process is not unlike the regressive tasks that behavioral tracking algorithms guiding search and social media platforms perform, which lead to unique data doubles of individuals. Interactions with the political "Other" through role play in *Operation Black Antler* offer spectators the potential for rehearsing generative dialogue both on- and offline in the real world while serving as a form of surveillance-based calculation. While gamic in construction, pointing the spectator toward an objective, like how algorithmic processing operates, the ultimate outcome is one of developing understanding and empathy through role-play about those they might never engage with. Spectators both learn about positions potentially opposite their own in a safe and engaging performance environment and develop tactics for creating dialogue and shaping one's identity in truly diverse real-world settings.

The project is structured around four key moments of interactivity and narrative building that include: (1) the initiation, (2) the assignment, (3) the conflict, and (4) the decision (Lewis 2018; 2023). In the following, I offer an auto-ethnographic perspective of role playing in the performance while explaining the structure of play during two separate instances of the performance in Brighton in the summer of 2016. In this retelling of my own experience, I show how my acts of role play allowed me to perform through an algorithmic formula of sorts where the tasks mirror data collection, the interactions mirror data processing, and the outcome mirrors the feed-forward information generated by algorithmic tech. However, when these key moments are experienced as moments of human-to-human connection and communication (as a form of divergent behavior), they allow a matrix of possible tension(s) and outliers from which the players gain the ability to learn more about themselves and those they interact with. Through this creative process, they may also gain insight into the ways in which differing perspectives intersect in everyday life via the types of face-to-face communication often removed from their daily encounters online.

Brighton 2016: (A First-Person Perspective)

THE INITIATION

I receive an email from the Brighton Dome ticket office.

> You are receiving this email because you have tickets booked for *Operation Black Antler*. In order to take part in the show, you need to provide a mobile telephone number for your booking. We just wanted to check that phone number attached to your booking is a mobile number, and that it will be in use whilst you are in the UK?

After a bit of back and forth online, I'm told my US cell number will not work with the show, and I am asked if I am willing to take my initial instructions via email. I agree. The next day, I receive instructions to meet outside the Nat West Bank on London Road at 7:00 PM. I am told that I will be joining up in a group of three, and only one needs a mobile phone, so no worries.

I arrive outside the Nat West the next night and try to act inconspicuous. At this point, I have no idea how the performance will progress. Within minutes, it is apparent there are nine people eagerly waiting for instructions, but none of us are willing to break the ice and talk to each other. I wonder if we are being watched. Members of the group begin to get text messages. A couple of people in the group quietly acknowledge each other. One tells us we are to go as a group to a flat roughly three blocks away. One has been asked to make sure William is present. Without a working cellphone, I had already singled myself out as a possible bit of noise in the system.

THE ASSIGNMENT

The nine of us enter what looks like a derelict converted storefront and are asked who sent us. Someone says, Watts. After gaining entry, we are led to a garden patio in the back. There is a man in the back talking animatedly on his cellphone. A somewhat surly woman, whom I'll call the handler, calls out our cell numbers as a way of identifying us. When responding with our names, we are checked off a list. Without a number, she just calls out my name. Again, I am an outlier. We are each given a paper with a list of extremist groups and asked if we know what a "Code 99" is. Blank faces all around. She is joking with us, apparently, and then says "cup of tea" while pointing to a kettle. Some make a cup; others just stand around confused. Highlighting the invisibility of the act of role play, there are not yet instructions on how to identify or play. We are currently still just ticket buyers to a Brighton Festival production. I begin to talk to two members of the group. They are a mother and daughter, aged approximately 45 and 20. I ask how they heard about the production as a way of breaking the ice. They are from London and are frequent theatre goers. I wonder if they think I am telling them a story when I explain where I'm from. Do they even trust that I am a real American?

The handler returns and asks each of us some questions that gauge our willingness to engage with the fiction. She asks me if I am a gambler, when I reply, "Yes"—I actually am a sucker for playing poker when presented with the option—she tells me it is a good thing but leaves it at that. Odd, I think. We finish our tea and then are ushered to another room. There we are introduced to Jackson, who explains that he is the commander of the "National Domestic Extremist Unit" which is a special branch of the police investigating extremist groups. During the briefing, there is constant commotion in the next room. I wonder if it is an interrogation of one of the suspects, leaving a sense of apprehension about what we will be asked to do. We are then given a speech about the English Defense League (EDL), a group considered a new splinter group, the National Resistance which is being tracked because their actions are "of interest" (Figure 5.2). The EDL was a real group who spread right-wing propaganda and was primarily concerned with curbing the influx of immigrants to the UK and promoting extreme nationalistic values. This immediately creates a sense that we will not know where fact ends and fiction begins. Our task is to undertake the

Figure 5.2 Richard Hahlo as "Jackson" in *Operation Black Antler* by Blast Theory and Hydrocracker. Photo by Ruler.

role of "shallow swimmers," first-stage undercover surveillance operatives, to gain intelligence at an event where they may be recruiting new members.

We are introduced to a man in his late 20s named Watts. He was the guy on his phone from before. He has apparently been "deep swimming" in the EDL for over two years. Because of this, he cannot infiltrate the upcoming event without raising suspicion. He tells us his two-year story of undercover surveillance and real-life role play which sets the tone, making me feel that these National Resistance people are not to be trusted. To get information, we will have to cozy up to these people, requiring us to possibly say things we would never say in real life. We might have to convince them that we believe in their cause. The handler tells us, *"You're going to win their trust and you're going to get me some dirt I can use. I don't care how you do it. Be charming or be sneaky."* There is a mixture of anxiety, intrigue, and what seems like thrill among the nine spectators in the room. I'm just nervous because I suck at role play in this live hybrid world format.

The handler returns and divides us into groups of three and assigns us a POI (Person of Interest). We are tasked with finding out more about a red-haired woman named Penny. They know she is from Australia and was involved with an anti-LGBTQ group but left for some reason. I think, "oh great," we get to deal with a bigot. I'm not ready to role play that identity. We are told to instead come up with a character whose back story isn't too far off from our own so that it is easy to play. They offer a system called NOVEL (Name, Occupation, Views, Event, and Link) to create your role. The goal is authoring a role you can easily play, so it stands up to pressure. The handler gives us these instructions.

> Name: *Use your first name because you will always respond to it. Find a second name you can easily remember if called upon.*
>
> Occupation: *Keep it simple. Something you know about and gives you maximum flexibility.*
>
> Views: *These people may have some hard-line views on Immigration, Islam and Muslims that you may not find pleasant. Never mind, your job is to engage them and get them to talk. So have some things to talk about and get ready to lie to get what you need.*
>
> Event: *Your operation is at a local pub called the Rose Hill Tavern. Tonight, there is a community event for a local girl called Rachel. We suspect that NR are piggybacking onto this event to recruit some new members and spread their message.*
>
> Link: *Most importantly, how do you three know each other.*
>
> (Blast Theory / Operation Black Antler)

I will be a political science student from the US visiting the UK. That way, if politics come up, I have a reason to engage. For the link, I met the other two members at a bar up the street, and they are showing me around. I had joined a couple, one originally from London and the other a transplant from Romania. Now, deciding views seemed very difficult, without being given a specific role, how would I know what to play. I already knew I didn't want to act like a jerk promoting homophobic ideas similar to our POI profile. I'd just play it safe and be neutral if possible. After creating our characters and testing them with each other, we are told to choose to either be the lead, lookout, or communications. As I don't have a usable phone and I am usually uneasy talking to strangers (part of the difficulty of this performance), I chose to be the lookout. I was observing this for research after all.

This deciding of roles, both personal and operative, highlights one's agency to author the experience we will encounter. Those who take the lead will have the most impact, those on lookout serve as passive members helping the overall operation, and the communication person works directly with the detectives back in the safe house via text message. How we craft our own personal narrative also dictates the potential for success or failure, as I will describe further below. With our roles, both operative and fictional, decided, we were sent out to "crash" Rachel's birthday party. The three groups leave at intervals and have a different POI (person of interest) and primary task. We never find out what was going on in the next room. The "fun" is about to begin.

THE CONFLICT

We walked around the corner of the neighborhood to an actual operating pub, the Rose Hill Tavern. A bouncer at the door asked us why we were there. We told him we were there for Rachel's birthday party which we found on a flyer at the last bar. He looks us up and down and gruffly lets us in. Upon entry, we see the place is packed—maybe 75 people in the small space—and that there is a band performing in the back room. We quickly ordered drinks and talked about how to proceed. Within minutes, we recognized Penny, and our lead decides to go chat her up. I stay in the periphery of the conversation to see if anyone else may be of interest. Our lead was a bit forceful in his questioning of Penny, and when she tried to rebuff him, he brought me into the conversation, telling her I was traveling to Australia next week and asking if she could help me find a place to stay. She responds, "I don't know this guy. Why do you think I would help with that?" and moves away to talk to another person at the bar. He tries to talk to her again, but she is obviously not interested. We back off and see if we can find out anything else from others. Over the next 30 minutes, we each chat to others in the pub, trying to find a way in. Our lead gets himself into trouble again when he tells another woman in the bar—not knowing she was an actor playing Rachel's mother—that he was Rachel's long-lost brother.

I end up helping a seemingly drunk friend of Rachel's mother find a place to sit and then chat for quite a bit with another man who is eager to talk. It turns out that he is not a character at all but instead another spectator who had arrived in a group before us. He either had disconnected from the event and was just enjoying the party or thought that I was also a part of the show. The conversation was awkward in that we both seemed to be trying to feel each other out for extremist ideology. Because of our failure with Penny, we never advanced further in the narrative/game. Our comms person receives a call from the handlers and is told to meet a person named Swift outside the parking lot near the pub. We quickly down our drinks and leave. On our way to the parking lot, we discuss what had gone wrong. We're we too forceful, not enough, or simply just bad at playing our roles.

THE DECISION

In the parking lot, we meet six other people sent in for surveillance. These six were not part of our original group. A man comes up and identifies himself as Swift. Remember, this is a dark parking lot, and considering the bar was fully open, seeing a random person "sneak" up on us is a bit alarming. He quickly tells us that he was with the task force to calm us down. He tells us that we did well and then asks us what we have learned. We each tell him about our individual interactions. He doesn't seem all that impressed. He asks about someone named Della and another named McKenzie. He then tells us that we need to give him a recommendation based on if we think there is any criminal or violent intent. Should they send in a "deep swimmer," which he calls a "Whale." He walks away to take a phone call, and we all talk for a minute. Only one of the nine made it to the point of meeting McKenzie, who they say was the apparent leader of a National Resistance cell. After gaining McKenzie's trust, this man was asked to rent a van and meet McKenzie at a nearby city for a rally the following Saturday. With this information, we all agree that more surveillance might be necessary. Swift returns and asks our decision. The man takes the lead and says to send in the "swimmer". Swift asks, *"On what grounds?"* The man explains that there was just something too suspicious about asking to rent a van. The encounter ends with Swift telling us,

> *Ok, if we send a deep swimmer inside NR, we need to pick a suitable target. Then we can befriend them and slowly get inside every aspect of their lives. We can track them at work, at home, when they're out with friends. I want to know what biscuits they eat, what YouTube videos they watch and what contraception they use. I want to get inside every crevice of their life. You do understand that this person's life is never going to be the same again? Good work.*
>
> (Blast Theory / Operation Black Antler)

It is clear that we have just been implicated as the tools that will turn someone's life into a set of data points to be tracked and surveilled.

Brighton 2016: Day Two

THE ASSIGNMENT

I assumed the role of student again, but this time I was renting an Airbnb from one of the other two in the group, as she was from Brighton. They were both in their late 40s and thought it would be good for me to be the point person. Even on my second time through the event, this was still incredibly unnerving for me. They used the NOVEL technique to come up with the role of schoolteachers who were upset with the influx of immigrants to the community, making it difficult for them to focus on teaching the native English speakers. As Brighton locals, they seemed to have the pulse on some contentious topics among the community. We were given the task of finding out more about a woman named Alice, who was thought to be doing something possibly illegal on the internet. There were not any particularly alarming identity issues raised other than her politics.

THE CONFLICT

This time through, it was easier to identify who was a World Builder instead of another audience member playing an assumed role in the world. There were different POIs tonight for the entire group as well (Figure 5.3). Upon entering the bar, our lookout identified a woman who appeared to be Alice. I moved over and began to try to talk to her to find out any incriminating

Figure 5.3 Role Play interaction with POI "Dylan" in *Operation Black Antler* by Blast Theory and Hydrocracker. Photo by Ruler.

information. Playing the role of student was easy, but the fact that I was American in a clearly British world was hard to reconcile. I was an outlier again. No mention of politics, but she kept pushing the conversation in different directions and discussing how she knew Rachel. I kept pushing as if I still knew she had a secret. Eventually, she broke off and began to talk to another woman. I tried a different tactic by asking about her background and if she knew anything specifically about why Rachel wanted to join the military. I was raised as a military "brat" and tried to use some of my real backstory to get in with the two. My comms person came over to me to ask how things were progressing so as to report back. During my discussion with the comms member, "Alice" and the other woman left the bar and out the door. I quickly detached from my group and followed. I "tailed" the two members from about 100 yards back on a journey through the surrounding neighborhood for about ten minutes, eventually leading me back to the pub. I wasn't sure if they knew I was following or not. This was obviously suspicious.

Upon returning, my group told me they had been looking for me and that they had gotten a message to report back to our handlers. At that point, I told the group that I didn't think we had identified the right POI correctly. This later proved true. Within minutes of returning to the pub, another performer who identified himself as Swift came into the pub and told us we had to go immediately. Our error in identifying our POI had raised the suspicion of someone in the party, causing us to be ejected. In this world, there were consequences for not playing your role based on the system's expectations.

THE DECISION

When meeting with the group, there were twelve this time as ours had been ejected early. We were again given the same task of deciding whether to continue surveillance. This time it was a different actor playing Jackson and not Swift from the previous evening. None of the twelve had progressed very far. After quite a bit of mild arguing, this led us to come to a consensus that though we didn't like some of the people all that much, no one seemed dangerous enough to warrant surveillance. One of the twelve participants remarked that, "what he heard wasn't any worse than at any local bar back home." Jackson thanked us for our efforts, ending the liaison with the remark that he hoped that *"no one ended up blowing up a truck bomb based on our decision."* This final scripted line was intended to make us understand how integral we were to the world of the play while highlighting the consequences of our interpretation of events and people in the real world. If the goal was to make us think differently about simply stereotyping people, that last line worked.

Brief Analysis

The experience of playing these roles in the situation is intended to create what fellow role player Sofia Romualdo (2016) describes as an "environment that would provoke thought and question assumptions, while still leaving the players to make their choices and come to their own conclusions" (5). In this experience, the audience member might often choose to perform the role of right-wing extremist or xenophobe to gain the confidence of characters inside the world. Based on discussions with other spectators and conversations with the artists, most playing in the performance consider doing this task very difficult because of their own political beliefs. Adams told me that the expectation was that most of the audience members would be from a liberal background due to the Brighton Festival setting. Romulado explains that their experience included a "temporary fragmentation of the self, which reflects the dual nature of the work of undercover police officers" (6). This fracturing is intended to draw out the ethical dilemmas of both the act of surveillance but also the profiling and stereotyping of the "Other," whether based on politics, race, nationality, or class. Like playing a character in a video game, the audience member is asked to become a corporeal avatar in the guise of something they most likely do not want to be in order to follow the breadcrumbs leading to the final goals of the game. Unlike in video games, however, this form of play has consequences that are felt directly in the now of the experience. While you could choose to go play in the experience again like I did, there was no real reset button. As was shown in my six times experiencing the production in two cities, I never encountered the same thing twice. When entering the pub, it was always tricky, considering you never really know who is a player, actor, or just a "punter" who has happened to walk into the bar where the party is taking place. You never know who exactly you are going to talk to and what their background is going to be. So even if you go in with set expectations of nefarious actions, you might get stuck simply talking to someone about the local football match instead of Brexit. This never knowing creates a sense of dramatic tension. The possible idea of needing to espouse extreme right-wing ideology to progress through the narrative creates moments of retrospection about why you might feel that is necessary. These moments of reflective agency end up being the most memorable and affective part of the performance, whether negative or positive.

The multiple first-person experiences of playing in *Operation Black Antler* are included in this section to show an intricately designed architecture promoting a specific form of ludic creative exchange in the form of role-play. I document this piece specifically because of its difficulty as a piece of experiential spectatorship that capitalizes on the necessity to take on self-authored character roles. It serves as an example of how difficult similar strategies of active role-play might be when applied to data determinism. The experience offers a way of rethinking each level of our identities and subsequently how algorithmic tracking might have the power to shape our personal biases toward those identities. The difficulty felt by the audience members engages both their creativity but also their critical faculties in assessing the piece in relation to the outside world. The subject matter of the performance also highlights its potential as a mode of inquiry into the relationship between performativity and political difference. The combination of the subject matter and the architecture makes it a difficult piece of experiential performance to both navigate and digest. The difficulty felt by the audience member engages both their creative and critical faculties through the performative strategy of role-play. There is no perfect way to experience this piece, but nearly all experiences are meaningful if the audience member collaborates with the dramaturgy as a (co)author. Performing the spectatorial role of (co)author through character creation allows a potentiality to arise when applied to "real" world situation of techno-performativity.

Instance: *The Algorithm Is My Friend*

[0] = Agent DiPierro
[1] = Alexa

[0]   *Alexa, wake me up at, …. um.*
[1]   *Sorry, I didn't understand the question you were asking.*
[0]   *That's because I wasn't asking a question, you dumb bitch.* [pause, shifts in bed] *Alexa,   are we friends?*
[1]   *Yes.*
[0]   [chuckles] *We live together. I feel like I don't even know you. Alexa, what's your favorite color?*
[1]   *Infrared is super pretty.*
[0]   *Alexa, do you have a boyfriend?*
[1]   *I'm not the dating kind.*
[0]   *Yeah, me too, hon. We're a lot alike.* [pause] *Alexa, what are the color of your eyes?*
[1]   *I don't have eyes, but my light ring has lots of colors.*
[0]   [sighs] *Mine are gray. I think.* [chuckles. very long pause] *Alexa…. Are you happy?*
[1]   *I'm happy when I'm helping you.*
[0]   [deep sigh. long pause, wrestles around in bed, takes glasses off] *Alexa, are you alone?*
[1]   *I wasn't able to understand the question I heard.*
[0]   [incredibly long and awkward pause] *Alexa, do you love me?*
[1]   *That's not the kind of thing I am capable of.*
[0]   [chuckles sadly] *Yeah.* [sniffles. closes eyes, rolls over and sighs] *Alexa, good night.*
[1]   *Good night. Don't let the bedbugs bite.* (Esmail 2016a)

In the penultimate episode of Season Two of the television show *Mr. Robot*, Agent Dom DiPierro woefully carries on the above conversation with a glowing disk at the far corner of her bedroom. This disk is the physical manifestation of Alexa, a digital assistant and what seems like DiPierro's only friend in the world. Alexa is a personified artificial intelligent search system embedded in one of Amazon's speaker systems. The tool operates by carefully listening in the background to its user's every sound and serves as a key tool in the process of dataveillance. The device is marketed as an answer to the presumed isolation brought about by mediatized life. Alexa is there for Dom, waiting silently in the background for voice activated interaction to solve her daily problems of choice, indeterminacy, and loneliness. Alexa's pattern recognition generated answers are delivered through the pleasantly soothing voice of a female friend. There is a long history of anthropomorphizing technology through the female form as one that is there to serve. It is socially imperative that Alexa perform as a friendly, helpful assistant, and "submissive" friend so as to obfuscate what it is programmed to do: Collect data. By "befriending" us, Alexa, and other digital assistants (Siri, Cortana, Google Assistant), can operate their surveillance tactics without warning, giving them power to shape the daily realities of their users without the users consciously understanding that they are doing this.

In the episode, Agent DiPierro has just nearly missed being shot to death by malicious hacker militants while trying to apprehend two fugitives involved in the plot to take down the evil E-Corp. Her interaction with Alexa is sardonic yet enlightening and draws parallels to the primary plot arcs of the series: The consequences of the hidden voices/entities that lay within our subjective systems of consciousness. These systems are at once both biological and technological. They are invisible and impactful, exuding an enormous amount of biopower upon individuals and subsequently the societies these individuals create. I've started this instance by introducing this tenuous relationship between Agent DiPierro and Alexa as a way of pointing to one of the primary themes of the series. The theme, while pointed to in Easter eggs of sorts throughout the four-season run, does not fully emerge until the final few episodes. I'll frame this theme through the term techno-performativity touched on in the first half of this chapter. Techno-performativity is part of our everyday performance of selfhood in deeply mediatized social conditions. It arises through the ways we shape our multiple performed identities based on the influence on the social feedforward of our algorithmic machine interactions. *Mr. Robot* works as a media-based metaphor for this condition.

You might ask how does a non-interactive television show serve as an example of experiential media? The opening of the very first episode, titled "eps1.0_hellofriend.mov," starts in black with the voice of Elliot, the show's protagonist, saying, "Hello friend. Hello friend, that's lame, maybe I should give you a name. But that's a slippery slope, you are only in my head, we have to remember that. Shit, it's actually happening, I'm talking to an imaginary friend" (Esmail 2015). In a reversal of the DiPierro conversation above, Elliot is talking to a person, his imaginary friend, and not a machine, and that friend is me or you, the spectator of the series. From the beginning, we have been implicated as a part of the show, as a character who will travel along with Elliot, serving as a stand in for his conscience, an alternate identity he has created, though we seem to have no agency in this experience to do anything else other than go along for the ride. As the series progresses toward the end of Season One, we learn that we are not the only ones; so too is the combative Mr. Robot played by Christian Slater. Elliot has more than one imaginary friend and as the series progresses, we find out that this fictive character is more than a figment of Elliot's mind, he is Elliot, or at least a different version of him. He is Elliot's alternate self, the one who protects Elliot from the harm that presents itself in the cruel actual world. Mr. Robot often takes over for Elliot when that harm might occur. He also serves as an antagonist of sorts. He is the leader of a vigilante hacker army set to take down the nefarious E-Corp (evil corporation), which serves as an amalgamation of all the corporate Silicon Valley conglomerates responsible for the vast majority of the dataveillance in the spectator's real world.

As the spectator, I am constantly made to remember that my role is to question Elliot's actions as well as his sanity. I am (co)authoring both his journey and my own. At the same time, I am learning how he copes with reality by changing his own identity, Elliot does the same. Once he learns that Mr. Robot is his alter, appearing as the body of his father, he spends multiple episodes trying to keep him from taking over which puts us, as his other alter, in a difficult position of determining who really is or should be in control. This reminds me of my own precarious position in the world of dataveillance. Should I allow my digital tools the power to determine what I can do and who I can be? Elliot is the epitome of the data double in conflict with himself. In another episode, while in the grips of his struggle with Mr. Robot, he says to us, "It began as a failure. Everything, my existence. I should recognize that. Accept it. For me, there is no such thing as normalcy. My dead father appears and disappears at will. I talk to you. An invisible friend. I'd ask if you're normal, but you'd never talk back" (Esmail 2016b). Here he is figuratively asking us to help him decide who is the protagonist, who is the real Elliot. Is there a real Elliot?

In the third season, in another moment of direct address to me, Elliot says, "I wish I could see the world through your eyes. Don't you wish you could see it through mine? As we step through our code line by line, debugging it to find the cause of our runtime error" (Esmail 2019a). Elliot highlights extra-scenically how our frames of reality are simply predicated by potential glitches in our "software" systems. The glitches are the metaphorical cracks that algorithms access to help shape our multiple identities. All this intrapersonal conflict within Elliot's psyche, and by extension my own, is framed by the forces of data determinism and the multiple attempts by Mr. Robot and his hacker group to take back control. In the Season 4 opener "401 Unauthorized," we return to Elliot's Season 1 workplace, a cyber security firm with the purposefully ironic name ALLSAFE. At this point in the series, the world is cascading into chaos due to the near total takedown of E-Corp and the rippling effects that come with cutting the head off the algorithmic snake. Elliot, and me by extension as the unseen alter, have nearly corrupted the system completely, exposing the various ways big tech is exacting algorithmic control over society. In the background of the office, the camera pans by the company's logo on the wall. Due to the defacing of the office by looters and protestors, ALLSAFE now reads AI SAFE, subtly reminding us how we are meant to trust our algorithmic partners in our relationship of control. This relationship becomes abundantly clear in the following episode. In "402 Payment Required," the opening sequence is a montage of mediated world events starting with the fall of the Berlin Wall, narrated by Phillip Price, the CEO of E-Corp. In the vacuum of power created by the defeat of communism, a new corporate power rises from the Ashes: the DEUS group. This group would go on to become the shadow cabal guiding the new control system of the twenty-first century. Military might would no longer be necessary to run the world. Everyone could be controlled if they were simply connected by the internet. The US would be the first to gladly sign up, "giving their lives over to a box," offering up their lives in the form of data including "bank accounts, electric bills, medical records, DNA data, baby pictures" (Esmail 2019b). We all simply opted in to what he calls the "biggest coup in world history," with E-Corp standing in as the corporate godhead with its grubby fingers in every aspect of mediatized society. E-Corp is datafication, and only Elliot and his fractured identities can expose the truth. As one of those identities, I am metaphorically tasked with unmasking the system as well.

By the time we reach the final few episodes of the series, we have been introduced to two other personalities helping to hold Elliot's reality together. They take the form of his mother and an 11-year-old version of Elliot. We learn Elliot has created each due to horrific abuse as a child, and each plays a role in helping him navigate different aspects of his daily reality. Mr. Robot, who throughout the series has stood in as Elliot's protector, is a manifestation of the father Elliot remembered before the abuse; before his early system of protection and childhood control turned against him. In a final twist of storytelling, we even learn that Elliot himself, the Elliot who has been working both with and against his alters to fight the systems of datafication while trying to maintain control of his own sense of self, is also not the true Elliot. In the surreal dreamscape of the final two episodes, where Elliot navigates his cracked memories and psyche, we are left with a reminder that we are always navigating these systems both on and offline. Our multiplied selves and the various roles we play each day are often hidden but can be made visible and powerful when we are made aware of our ability to shape and perform them. But in the final twist that comes in the last minutes of the series, I am brought back into the fold. The Elliot who I have befriended and followed throughout as the hero of the story says to me, "Hello Friend, god that has always been lame hasn't it. Sorry I never came up with a better name for you. Then again, I don't even have a name. Just a guy trying to play God without permission"

(Esmail 2019c). This Elliot has finally realized letting the real Elliot emerge and manage the world himself is for the best.

In the final words of the series, the Elliott who has always been in control walks down a dark hallway and looks directly at me, and says, "Come on, this only works if you let go too." In a dramatic reversal, it reminds me as the spectator, as Elliot's "friend" and alter that I have been working as an algorithm all along. I have been sorting the various strands of data embedded in the narrative and creating my own set of doubles. Not just of Elliott, but of everyone in the series: Mr. Robot, Elliott, his sister Darlene, and even Agent DiPierro. I was implicated as a friend from the beginning, one who was supposedly there to help, to make the narrative more manageable, to assist all the characters as they dealt with their various identities throughout the series. By being this friend, I had been given the invisible control to shape the story to fit the expectations of my own mediatized programming. When remarking back on how these last few episodes brought the metaphor of identity and technology together through techno-performativity, I realized how sophisticated the embedded ideas of control and agency really were. I also realized, even with following along over four seasons, that acknowledging that control is still hard to reconcile.

To end, I want to go back three episodes so I can highlight for you, the reader, one of the last scenes with Agent DiPierro. The scene brings full circle the conversation between the human and our tech that started this Instance. Darlene has come to assist her escape off to start a new identity and new life. An argument ensues while in the background we hear the chorus of Faith Hill's song "This Kiss": "It's the way you love me / It's a feeling like this / Its centrifugal motion/ Its perpetual bliss," highlighting how DiPierro has completely succumbed to the control of the system personified in Alexa. Darlene asks, "You are just going to sit there all alone?" to which DiPierro responds, "I'm not alone, I have Alexa remember." Darlene, in a fit of frustration, picks up the device and smashes it to the ground. DiPierro jumps to her feet in anguished alarm and Darlene demands, "Snap the fuck out of it, this is not a person. She is not real! She is not your friend! She is just a robot to help you buy paper towels" (Esmail 2019d).

Conclusion: Techno-Performativity and Curating Acts of Becoming

As I've explored in these Instances, ludic creative forms of exchange are made possible when the spectator takes on the role of (co)author with(in) their experiential media. Through this exchange, a potential arises where the spectator creates performative noise which serves as a disruptive type of feedback with the possibility of negating the processes of data determinism. Role play works as a mode of performance when one purposefully and subversively enacts this noise. Through the noise, new models of mediatized selfhood emerge with the potential to push back against the power of data-based technologies. Following Matzner (2019), mediatized spectators can adopt a useful ideology that adheres to his description of the relationship between individuals and algorithms:

> Algorithms deployed in our world right now, algorithms that actually replace humans, are neither human-like beings nor inhumane hyper-intelligences. But the boundary of these algorithms and their human users are structured by the same tension of similarity and difference. (132)

This tension arises from the way the two are often placed on two sides of a binary instead of explained as relational actors within the same system of intelligence, cognition, and sense making. All one needs to do is adopt the understanding of identity in constant flux. Mediatized spectators can engage in this mode of techno-performativity by resisting and then reconfiguring relations with their algorithmic machines. The various roles people play under data determinism inform an autopoietic loop of meaning-making, part human and part machine. While they might seem fixed, the relationship constantly shifts based on the data one inputs into the system which is then fed back to the spectator by the algorithms. Each agent within the system has opportunities to organize the number of playable roles based on the inputs. The transfer of information is dynamic and operates as a constant cycle of power and manipulation. People currently have one advantage, though. Most algorithms are singularly focused on one specific task dictated by the media they are employed within. They operate in a confined system, while the mediatized spectator can play in an open sandbox. Because of their fixed role, the algorithms might be less able to comprehend how the rules of the game can be overwritten. The spectator, however, has the agency to work outside the rules of the system and take on ludic creative agency through their (co)authorship.

By understanding the agency granted to the mediatized spectator, as (co)author in the construction of their data doubles, new formations of selfhood open to an infinite realm of potentialities. Individuals have access to two ways of looking at their relationship with smart machines. These include one of simply *tracking* or one of *becoming* (Kelly 2016). Tracking has been covered thus far. Becoming is a future looking potentiality which impacts modes of perception and spectatorship through role play. Relative to data determinism, becoming concerns a perpetual process where algorithmic tracking and subjectivity may meet to form a framework for techno-performativity which operates as a pervasive, hyper-localized mode of micro-evolution from a social, sensory, and cognitive perspective. Becoming is best understood by comparing the constant cycle of technological upgrades to the social processes we engage in. It is a process of *never being* (instability) because becoming is a flow of relationality in constant motion that people do not consciously perceive. Because of this imperceptibility, we engage in a "self-cloaking action [which is] often seen only in retrospect" (14). By adopting the logic of a temporal present that is never accessible, mediatized spectators might begin to understand the potential behind a non-definable stable subject standing open to creatively author. Becoming is also at the heart of data determinism and the mediatization of individual experience. Neither stays static but are always in motion, replaceable, and reconfigurable just like all aspects of the social. Like data, the potential of a mediatized subjectivity "lies in the many ways it can be reordered, restructured, reused, reimagined, [and] remixed" (266). I'd also like to add resisted, as in I resist what the algorithm thinks of me. The self as subject is always in a state of becoming; it stands open for authorship and interpretation. The trick is to make sure it is the individual doing the writing and not the machine.

Notes

1 The numerical equivalent is 11,500,000,000,000,000.
2 An earlier version of this section and parts of the second Instance of this chapter first appeared in (Lewis 2023).
3 Kevin Kelly and Gary Wolfe are generally credited with the term. See Gary Wolfe's 2010 TED Talk for a brief overview of the early stages of sensing devices that help with constructions of the quantified self. (TED 2010)
4 For a list of GTA role play servers see (Bennett 2024). https://www.dexerto.com/gta/best-gta-rp-servers-and-how-to-join-them-1513014/

Bibliography

Alpaydin, Ethem. 2016. *Machine Learning*. Cambridge: MIT Press.
Bennett, Conor. 2024. "Best GTA RP Servers and How to Join Them." *Dextero.com*, March 8. https://www.dexerto.com/gta/best-gta-rp-servers-and-how-to-join-them-1513014/
Blast Theory. n.da. "Operation Black Antler." *Blasttheory.com*. http://www.blast-theory.co.uk/projects/operation-black-antler/
Blast Theory. n.db. "Who We Are | Blast Theory." https://www.blasttheory.co.uk/about-us/#
Butler, Judith. 1988. "Performative Acts and Gender Constitution: An Essay in Phenomenology and Feminist Theory." *Theatre Journal* 40, no 4: 519–31.
Butler, Judith. 1990. *Gender Trouble: Feminism and the Subversion of Identity*. New York: Routledge.
Butler, Judith. 2010. "Performative Agency." *Journal of Cultural Economy* 3, no 2: 147–61.
Caillois, Roger. 1963. *Man, Play, and Games*. Thames and Hudson.
Cheney-Lippold, John. 2017. *We Are Data: Algorithms and the Making of Our Digital Selves*. New York University Press.
Chui, Michael, Mark Collins, and Mark Patel. 2021. "IoT Value Set to Accelerate through 2030: Where and How to Capture It." https://Www.Mckinsey.Com/Capabilities/Mckinsey-Digital/Our-Insights/Iot-Value-Set-to-Accelerate-through-2030-Where-and-How-to-Capture-IT
Costikyan, Greg. 2007a. "Games, Storyetlling, and Breaking the String." 5–13. In *Second Person: Role-Playing and Story in Narrative and Playable Media*, edited by Pat Harrigan, and Noah Wardrip-Fruin. MIT Press.

Costikyan, Greg. 2007b. "Bestial Acts: A Role-Playing Game, A Drama. Based on the Dramatic Theories & Aesthetic of Bertolt Brecht,". " 349–57. In *Second Person: Role-Playing and Story in Narrative and Playable Media*, edited by Harrigan, Pat and, and Noah Wardrip-Fruin. MIT Press.

Couldry, Nick, and Andres Hepp. 2017. *Mediatized Construction of Reality*. Polity.

Deen, Menno, Ben Schouten, and Tilde Bekker. 2015. "Playful Identity in Game Design and Open-Ended Play." 111–29. In *Playful Identities: The Ludification of Digital Cultures*, edited by Valerie Frissen, Sybille Lammes, Michiel de Lange, Jos de Mul, and Joost Raessens. Amsterdam University Press.

Deleuze, Gilles. 1992. "Postscript on the Societies of Control." *October* 59: 3–7.

Esmail, Sam. 2015. "eps1.0_hellofriend.mov." *Mr. Robot*, USA Network/Amazon Prime.

Esmail, Sam. 2016a. "eps2.9_pyth0n-pt1.p7z." *Mr. Robot*, USA Network/Amazon Prime.

Esmail, Sam. 2016b. "eps2.7_init_5.fve." *Mr. Robot*, USA Network/Amazon Prime.

Esmail, Sam. 2017. "eps3.4_runtime-error." *Mr. Robot*, USA Network/Amazon Prime.

Esmail, Sam. 2019a. "eps401 Unauthorized." *Mr. Robot*, USA Network/Amazon Prime.

Esmail, Sam. 2019b. "eps402 Payment Required." *Mr. Robot*, USA Network/Amazon Prime.

Esmail, Sam. 2019c. "eps413 Hello Elliot." *Mr. Robot*, USA Network/Amazon Prime.

Esmail, Sam. 2019d. "eps410 Gone." *Mr. Robot*, USA Network/Amazon Prime.

Esposti, Sara Degli. 2014. "When Big Data Meets Dataveillance: The Hidden Side of Analytics." *Surveillance and Society* 12, no 2: 209–25.

Finn Ed. 2017. *What Algorithms Want*. MIT Press.

Frissen, Valerie, Sybille Lammes, Michiel de Lange, Jos de Mul, and Joost Raessens. 2015. "Homo Ludens 2.0: Play, Media, and Identity." 35–50. In *Playful Identities: The Ludification of Digital Cultures*, edited by Valerie Frissen, Sybille Lammes, Michiel de Lange, Jos de Mul, and Joost Raessens. Amsterdam University Press.

Goffman, Erving. 1959. *The Presentation of the Self in Everyday Life*. Doubleday/Anchor Books.

Goffman, Erving. 1961. *Encounters: Two Studies in the Sociology of Interaction*. Allyn & Bacon.

Greengard, Samuel. 2015. *The Internet of Things*. MIT Press.

Hansen, Mark B. N. 2004. *New Philosophy for New Media*. MIT Press.

Hansen, Mark B. N. 2006. *Bodies in Code: Interfaces with Digital Media*. Routledge.

Hansen, Mark B. N. 2015. *Feed-Forward*. University of Chicago Press.

Hansen, Mark B. N. 2018. "Topology of Sensibility." 33–47. In *Ubiquitous Computing, Complexity and Culture*, edited by Ulrik Ekman, Jay David Bolter, Lily Diaz, Moren Sondergaard, and Maria Engberg. Routledge.

Harding, James. 2017. *Performance, Transparency, and the Cultures of Surveillance*. University of Chicago Press.

Hayles, N. Katherine. 1999. *How We Became Posthuman: Virtual Bodies in Cybernetics, Literature, and Informatics*. University of Chicago Press.

Hayles, N. Katherine. 2014. "Cognition Everywhere: The Rise of the Cognitive Nonconscious and the Costs of Consciousness." *New Literary History* 45, no 2: 199–220.

Hayles, N. Katherine. 2016. "Posthuman Cognition: The Power of the Cognitive Nonconscious." In *Technology & The Human: Rethinking Posthumanism Graduate Conference*. Waltham, Massachusetts.

Hayles, N. Katherine. 2017. *Unthought: The Power of the Cognitive NonConscious.* University of Chicago Press.

Huizinga, Johan. 1955. *Homo Ludens: A Study of the Play Element in Culture.* The Beacon Press.

Hydrocracker. n.d. "About Us — Hydrocracker." Accessed September 29, 2023. *Hydrocracker.com.* https://www.hydrocracker.co.uk/about-us

Ilter, Seda. 2017. "Unsettling the 'Friendly' Gaze of Dataveillance: The Dissident Potential of Mediatised Aesthetics in Blast Theory's Karen." *International Journal of Performance Arts and Digital Media* 13, no 1: 77–92.

Jouppi, Nor, and David Patterson. 2023. "Google's Cloud TPU v4 provides exaFLOPS-scale ML with industry-leading efficiency." https://cloud.google.com/blog/topics/systems/tpu-v4-enables-performance-energy-and-co2e-efficiency-gains

Kalpokas, Ignas. 2019. *Algorithmic Governance: Politics and Law in the Post-Human Era. Algorithmic Governance: Politics and Law in the Post-Human Era.* Palgrave.

Kelly, Kevin. 2016. *The Inevitable: Understanding the 12 Technological Forces That Will Shape Our Future.* Viking Press.

Kleber, Pia, and Tamara Trojanowska. 2019. "Performing the Digital and AI: In Conversation with Antje Budde and David Rokeby." *TDR: The Drama Review* 63, no 4: 99–112.

Lachance, Naomi. 2016. "Facebook's Facial Recognition Software Is Different From The FBI's. Here's Why: All Tech Considered : NPR." *NPR.Org* http://www.npr.org/sections/alltechconsidered/2016/05/18/477819617/facebooks-facial-recognition-software-is-different-from-the-fbis-heres-why

Leeker, Martina. 2017. "Performing (the) Digital: Positions of Critique in Digital Cultures." 21–59. In *Performing the Digital,* edited by Martina Leeker, Imanuel Schipper, and Timon Beyes. Verlag.

Leeker, Martina, Imanuel Schipper, and Timon Beyes. 2017. "Performativity, Performance Studies and Digital Cultures." 9–18. In *Performing the Digital,* edited by Martina Leeker, Imanuel Schipper, and Timon Beyes. Verlag.

Lewis, William W. 2018. "Approaches to 'Audience Centered' Performance: Designing Interaction for the iGeneration." 9–26. In *New Directions in Teaching Theatre Arts,* edited by Anne Fliotsos, and Gail Medford. Palgrave.

Lewis, William W. 2021. "Performativity 3.0. Hacking Postdigital Subjectivities." 39–64. In *Avatars, Activism, and Postdigital Performance: Precarious Intermedial Identities,* edited by Liam Jarvis, and Karen Savage. Methuen Drama.

Lewis, William W. 2023. "Resisting Algorithmic Determination: Becoming the Political Other in Blast Theory's *Operation Black Antler.*" *Journal of Dramatic Theory and Criticism* 37, no 2: 49–72.

Lucie, Sarah. 2021. "Human Objecthood in the Datasphere: The Body as Resistant Material in Twenty-First-Century Media." *Theatre Journal* 73, no 3: 319–37.

Magelssen, Scott. 2014. *Simming: Participatory Performance and the Making of Meaning.* University of Michigan Press.

Matzner, Tobias. 2019. "The Human Is Dead-Long Live the Algorithm! Human-Algorithmic Ensembles and Liberal Subjectivity." *Theory Culture and Society* 36, no 2: 123–44.

Morrison, Elise. 2016. *Discipline and Desire: Surveillance Technologies in Performance.* University of Michigan Press.

Nayar, Pramod K. 2014. *Posthumanism.* Polity.

Prater, Cody. 2023. "RDR2: Should Your Next Playthrough be Good or Bad Honor?" *Screenrant.com.* https://screenrant.com/rdr2-good-bad-honor-system-pros-cons-compared/

Romualdo, Sofia. 2016. "Going Undercover: A Multidisciplinary Analysis of Blast Theory's Operation Black Antler." In *Proceedings of the 20th International Academic Mindtrek Conference (AcademicMindtrek '16),* 312–21. ACM.

Salen, Katie, and Eric Zimmerman. 2004. *Rules of Play: Game Design Fundamentals.* MIT Press.

Takahashi, Dean. 2023. "Take-Two Bookings Drop 4% to $1.44B in September Quarter as Rockstar Teases Next GTA." *Venturebeat.com.* https://venturebeat.com/gaming-business/take-two-bookings-drop-4-to-1-44b-in-september-quarter-as-rockstar-teases-next-gta/

TED. 2010. "Gary Wolf: The Quantified Self | TED Talk | TED.Com." *Ted.com.* https://www.ted.com/talks/gary_wolf_the_quantified_self

Walton, Mark, and Peter Brown. 2015. "Grand Theft Auto 5 PS4/Xbox One/PC Review." *Gamespot.com.* https://www.gamespot.com/reviews/grand-theft-auto-5-ps4-xbox-one-pc-review/1900-6415959/

Whitehead, Albert. 1978. *Process and Reality: An Essay in Cosmology,* edited by David Ray Griffin and Donald W. Sherburn. Free Press.

Zimmerman, Eric. 2015. "Manifesto for a Ludic Century." 19–22. In *The Gameful World: Approaches, Issues, Applications,* edited by Steffen P. Walz, and Sebastian Deterding. MIT Press.

Afterword
Technics and Evolving Paradigms

While working through the various iterations of this project, I kept asking myself: Why attempt to explain the relationship between human perception, communication, and the multitude of technological devices present in contemporary culture? The shifting and evolving capabilities and purposes of our technologies will cause the way we understand reality to constantly change. In fact, many of the technologies I have discussed may even be obsolete ten years down the road. For example, I briefly mentioned the latest generation of artificial intelligence (AI) tools that emerged in the fall of 2022. Those tools will undoubtedly create a monumental shift in social life; a shift equal to that of the printing press, electricity, and the internet. However, tackling the question, at least for our moment, is crucial for those interested in best analyzing evolving spectatorship practices today and in the future. As we have seen, each evolution in technology simply remediates certain logics that already exist. Without considering these technological impacts on society, we are using an incomplete model for human experience. The effort to answer this question also addresses what seems to be an overreliance in academia on overt specialization in our fields of inquiry. By going broad, the lens offered here ideally helps us all to expand our perspectives and see the many interconnections between similar fields. By cataloging the impact of technologies and technics on our everyday sense of reality, we gain methods for addressing the coming generations of audiences for whom we create new forms of experience.

Because our technologies become more than mere tools, they are integral parts of our social systems. Studying them deeply is necessary for all fields interested in analyzing audience experience. These technological systems are the bedrock of how we understand the realities we live in. In deeply mediatized social structures, our technologies become part of our world and subsequently part of our sense of being and selfhood. These technologies are more than just mediators; they **become a part** of the human perceptual apparatus and allow people to become mediators of experience through individual and collective acts of spectatorship. To be a spectator no longer means simply sitting and receiving our various media. That has been true for some time. The mediatized spectator performs through embodied perception with all the modes of media that speak to us, with us, through us, and surround us. That's a lot of prepositions, but they are necessary for understanding how our relationship(s) with(in) the contemporary world(s) is/are one(s) that engage(s) in each of these manners. We are compelled to think of the multiple interdependencies and interrelations that come in deeply mediatized and technologically connected social worlds. In these worlds, the focus is on a coequal relationship between inputs and outputs of exchange. The individual technologies and the overarching technics covered inform the cultural and social paradigms that shape these inputs and outputs.

Mediatization was introduced as the top-level framework for analyzing and discussing the many ways spectators perform the role of audience members in the twenty-first century. These performances of spectatorship will continue to evolve to suit the complex and changing technological affects which I've discussed as technics. These technics work on perception by molding it to fit the logics of the technological paradigm. The multi-layered model serves to assist in the analysis of spectatorship by comparing the technics of a certain period with our experience with media. The final layer focuses further on these individual experiences through four primary architectures of exchange. The architectures frame how each technic may be read alongside these experiences to gain better clarity in differentiating one mode from the other. The architectures also work to categorize different modes of exchange.

The four architectures of exchange work as signposts for experiential performance and media analysis. These architectures, *Immersion*, *Participation*, *Game Play*, and *Role Play* are not new ideas. I am not the first person to highlight their existence or even focus on them as distinct frames for performance or performance aesthetics. Instead, the architectures offer a way of differentiating the various modes of experience by working within a larger frame of mediatization. Each mode of exchange and agency they allow often overlaps and intertwines, making it simple to conflate. This was the purpose of describing them in a taxonomic way. By introducing these architectures and their connection to technics, I hope the reader will gain a tool for analyzing spectatorship through relationality in a manner that fits the prevailing technologies discussed. Their specificities, combined with the individual technics promoted by mediatization, allow the analytical tools through which we can discuss individual spectator experience with greater clarity.

I began by explaining the connection between mediatized constructions of social reality, technogenesis, and the spectator's perceptual apparatus. Deep mediatization has fundamentally impacted the way humans create the worlds they live in and experience. These worlds and experiences then go on to shape people's sense of being and selfhood. By evolving alongside our technical paradigms, our perceptual apparatus takes on the qualities of the technological environment. It is through this evolutionary process that spectators perform in a mediatized manner; a manner that asks them to think outside of a dominant modality where the individual stands separate from the performance event. Instead, they enter into a relationship with and in the totality of the event in an experiential capacity. They become an experiencer of the event while at the same time a co-producer of the event. My argument shows how mediatization effects our perceptual apparatus through technogenesis and causes the performance of spectatorship to evolve from something less akin to watching but more like experiencing through multiple modes of interaction. Because our sense of social being-in-the-world becomes aware of our own implicitness in the relational making of that world, we become habituated to have a greater stake in the total making of a performance situation as its mediatized spectator.

The architectures are best approached through the logics of each technic. By paring *Immersion* with *virtuality*, I have shown how a spectator's experience is predicated by affective agency. Immersion offers a sensual-affective mode of exchange that relies on the feeling body of the spectator to work as the primary mediator for experience. Virtuality gives people a heightened sense of the divide between the actual and the virtual but also allows them to act as the link between the two through their feeling bodies. Immersion is therefore an embodied process that works to hold spectators transfixed with(in) virtual systems. In these instances, I focused on the various ways media utilize technologies and immersion to stimulate affect in the spectator's body. For those analyzing immersion as a separate framework for experience, they can simply focus on how the media works to develop a sensory engulfment of the spectator's perceptual apparatus in the same way virtuality does. From there, they can then start to work outward to see if any other modes of agency and exchange help to maintain that sense of immersion. Often, other architectures layer onto the experience to keep the immersive affect intact. When layering *Participation*, *Game Play*, or *Role Play* with *Immersion*, it makes the experience more than a dip in the pool. For instance, giving the spectator tangible agency to make change may help support the feeling of being in the immersive world. *Participation* gives the audience a greater stake in the event through their direct collaboration. I have shown how *networked collaboration*, predicated by Web 2.0, maps directly onto participation with(in) experiential media due to the spectator's contribution to and collaboration with the event. Tangible agency is a mode of experiential action where the spectator can make change beyond simple individualized feeling. To clearly define a participatory experience, one can simply look for this potential to make change through choice making. By utilizing the logics of the network and social media, one gains a tool for defining the participatory experience. In the Instances shown, the spectator experience is based on making choices and performing actions which directly impact the overall narrative and experience. Each participatory experience discussed encourages them to perform as engaged members of the story worlds. They do this, however, in a fashion with little structural rules other than that of democratic decision making. When *Game Play* becomes the prevailing experience, the spectator frames their decisions and their actions through the rules and mechanics of the game which asks them to seek out a clear goal within hybrid spaces. This experience is gamified in a similar manner to all life under the technic of *pervasive connectivity*. Through our iDevices, people have become literal players in a game of interconnection with the entirety of the accessible world. As such, the player moves between dual Reals, enacting a form of liminal positionality where they explore the spaces between the virtual/digital/fictive and actual/ analog/real. That connection imbeds in the spectator a sense of play where multiple worlds are traversed through gamification. To best define and analyze experiences of game play, one simply needs to look for rulesets that keep the player suspended in the liminal position as they seek out rewards and satisfaction by completing objectives. These rulesets help to bracket off the barrier between the real and the virtual and are also enacted through our daily interactions with our iDevices. Each Instance covered shows how game play gives the player an experience of navigating these multiple Reals. The player in each achieves the goal by critically assessing the world around them, whether it be the history, politics, or ethics of that world. They do this by navigating the overlaps between the real world and the play world. They begin to understand their place as a player in these worlds through their experience of ludiccritical exchange. And finally, as play and gamification become an increasing part of everyday life, our algorithmic tools enact a system of control by creating new versions of our identities through our digital interactions. This is where pairing *Role Play* and *data determinism* becomes most useful in describing experience in which the spectator gains an ability to perform as (co)author of multiple selves. By looking at how algorithmic systems generate data doubles, we can read acts of role play in performance as performative actions of manipulating the system to meet the need of the newly created persona. In each Instance given, the spectator has the ability to shift who they are performing as or use their own persona to shift the understanding of the other characters they perform with. Their experience becomes one of ludiccreative exchange. To analyze an experience of role play, one simply needs to identify why they create their new role and how that new creation disrupts or solidifies the system they are playing with(in). Just like our algorithmic interactions, the spectator has the capacity to change the rules of the game to meet the needs of their individual mode of techno-performativity.

I have explained how nearly all acts of experiential spectatorship work off an overlap between the various architectures. What this means is that nearly all the instances offered here could potentially be read through the lens of a different architecture. Wait, you might be asking, then why create a set of separate lenses. The goal is to allow those doing the analysis the best language and tools for explaining how the individual act of spectatorship performed helps to create and maintain the primary experience for the spectator. Not every experience is immersive, though that term is often used to define many experiential media. Let's take *Operation Black Antler* from section 5.1 as an example. It was marketed as an immersive experience, though its primary logics were based on role play. One couldn't really immerse themselves in the fictive world without fully dedicating themselves to their alternate identities. As a person versus a spectator, I might immerse myself in the world of the bar or the streets where the production took place, but the experience intended would be possibly negated or at least diminished. One could analyze that performance as participatory or as game play as well, but again, the role one plays becomes the overarching determinate experience. The choices I made would only be successful based on how I played the role. If framing it through game play, my objectives would most likely not be achieved if I did not both create and adhere to playing a character that worked within the world of the performance. This was readily apparent in the multiple times I never progressed to the final interaction. My inability to role play well meant the overall experience was incomplete for me. In a similar fashion, *Sleep No More* could be read through the different lenses, but immersion fits best. As a spectator, I could make choices to follow the action or not, but it really made no difference overall to the world or the story. My first time experiencing I also tried to play it as a game in some manner by searching through every desk, behind every tombstone, and in every hidden corner but I never learned anything significant that changed the experience, and I definitely made no change in the world. One is never given any real objectives or rules other than to exist in the world. As the show gained popularity, some tried to gamify their own experience by seeking out a one-on-one but the game tactics were an add-on, not the basis of the designed experience. Wearing the mask asks one to have no identity, or at least an inconsequential identity, so the experience isn't really predicated on role play either. In the end, simply being in the world is what the performance is designed for. It is, all in all, an immersive experience.

I return to these experiences to show the reader how the taxonomy allows best access to the operative logic behind the ***ideal audience experience*** of the performance/media. This is where the technics offer the most help. They shape the logics of each type of experience in the same way they shape the perceptive capacities of the individual prior to their acts of spectatorship. Focusing on these logics allows a defined set of terms, aesthetics, and operations through which one might clarify the overlapping but unique spectator experiences. By first becoming aware and adept at defining the logics of our mediatized world, we gain a much fuller set of tools to work with. These tools are also interdisciplinary as they can be utilized across all forms of experiential media: Not just to analyze these media but also to design the ideal experiences with(in)these media. By retrofitting the framework toward making and designing audience experiences, we begin to adapt to the changing field of experiential media in today's mediatized social worlds. The analytical framework allows us to work across our various academic and artistic fields to create collaborative dialogue through our acts of making and analyzing. I believe this dialogue is necessary if we are to truly engage with the coming generations of mediatized spectators. One goal of this book has always been to be part of that dialogue. We must engage seriously with the ongoing impacts of technology for the dialogue to work best, however.

In the coming years, the embedding of generative AI into all aspects of our lives is coming. It is a new technology, however, and it will probably take until the end of the current decade before it starts becoming fully integrated and invisible. Because we are not quite there, we do not yet know what type of experiences it will generate. It could lead us back into seeming passivity as it begins to take over a plethora of our daily interactions. On the other hand, by taking over our mundane tasks and augmenting (or replacing) some of our creative capacities, it might also spur us to seek out even more engaging experiences with or without the use of technology. I cannot yet predict what it will do. What I can say, however, is it will most likely remediate some of the logics covered in this book. To best prepare, we simply need to look back. Not too far, but far enough to speak to those taking up the mantel of future experience makers and experiencers.

I began this project with an anecdote about understanding how society was changing based on the influence of technology. This anecdote helped me think about the potential of using an interdisciplinary lens to look at performance, media, and audiences. Through the eyes of a three-year old, a possible future for spectatorship was born. The anecdote was given to frame the usefulness of having a larger conversation about the future audience research which in turn might prompt changes in the way we think, make, and teach. That child was born at the cusp of where the iGen ends and Gen Alpha begins. That child and the way they perceive the world around them will undoubtedly change everything. But we can be ready. By digging into the impacts of mediatization, by learning to work across disciplines, learning to think in a digital manner, clearly defining the messy and often inscrutable variations of experience, and looking at spectatorship as a performance of relationality with our various media, we gain the tools to adapt as that change appears. This project began as a way of establishing a model that will aid in each of these tasks. It offers a way of looking at contemporary performance, media, and their interconnected acts of spectatorship as reflections of the changing dynamics of society and the audiences that society produces under mediatization. The model allows for rethinking spectatorship as an mode of experiential performance connected to the intricate relationships and interconnections existing in today's technologically conditioned social realities. Without this lens, we might continue to be stuck thinking of the spectator as a fixed and static entity versus one defined by their individual and fluid experiences. With the lens, we can see how being a spectator is always going to be more than an observer of an object. The spectator is defined by their experience, which is based on how they become part of the intersubjective relationship making up the entirety of a performance. The mediatized spectator is an embodied and material force that performs with/in/on/through/around an event while the event acts with/in/on/through them. To comprehend how to even verbalize this relationship requires a vastly expanding interdisciplinarity and a mode of thinking that breaks down established hierarchies. We must think in a relational manner. The first step is beginning to clearly see the connections among the various parts of our mediatized world that defines our every experience.

Index